The Road Taken: An Automotive Saga

by

Etienne Psaila

The Road Taken: An Automotive Saga

Copyright © 2024 by Etienne Psaila. All rights reserved.

First Edition: **May 2024**

No part of this publication may be reproduced, distributed, or transmitted in any form or by any means, including photocopying, recording, or other electronic or mechanical methods, without the prior written permission of the publisher, except in the case of brief quotations embodied in critical reviews and certain other non-commercial uses permitted by copyright law.

This book is part of the 'Automotive and Motorcycle Books' series and each volume in the series is crafted with respect for the automotive and motorcycle brands discussed, utilizing brand names and related materials under the principles of fair use for educational purposes. The aim is to celebrate and inform, providing readers with a deeper appreciation for the engineering marvels and historical significance of these iconic brands.

Cover design by Etienne Psaila
Interior layout by Etienne Psaila

Website: **www.etiennepsaila.com**
Contact: **etipsaila@gmail.com**

Table of Contents

Part I: The Dawn of the Automobile
1. Chapter 1: A New Era Begins
2. Chapter 2: Building Dreams
3. Chapter 3: On the Open Road

Part II: The Roaring Twenties and Beyond
4. Chapter 4: The Golden Age of Automobiles
5. Chapter 5: Surviving the Great Depression
6. Chapter 6: World War II and the Automotive Shift

Part III: The Mid-Century Transformation
7. Chapter 7: The Rise of Muscle Cars
8. Chapter 8: Design and Innovation
9. Chapter 9: Road Trips and Cultural Shifts

Part IV: The Modern Era
10. Chapter 10: The Digital Revolution
11. Chapter 11: The Green Movement
12. Chapter 12: Autonomous and Connected Vehicles

Part V: Legacy and Future
13. Chapter 13: Reflecting on a Century of Change
14. Chapter 14: Passing the Torch
15. Chapter 15: A New Road Ahead

Epilogue: A Legacy of Innovation

Chapter 1: A New Era Begins

The Birth of Mass Production

In the early 20th century, the world stood on the brink of transformation. Cities buzzed with the hum of factories, and the air was thick with the promise of innovation. At the heart of this technological revolution was the automobile, a marvel that would redefine not just transportation but society itself.

Among the myriad of visionaries who saw potential in this burgeoning industry, none were as influential as Henry Ford. His creation, the Model T, wasn't just a car; it was the harbinger of a new era. Launched in 1908, the Model T was designed to be affordable, durable, and simple to operate. Ford's genius lay not only in the vehicle's design but in his revolutionary approach to production.

The assembly line, a concept borrowed from the meatpacking industry, was Ford's masterstroke. By breaking down the manufacturing process into simple, repeatable tasks, he was able to dramatically increase efficiency and lower costs. This method turned car production from a laborious, artisan craft into an industrial powerhouse. The Model T, affectionately dubbed the "Tin Lizzie," rolled off the assembly lines in numbers previously unimaginable. By 1927, Ford had produced over 15 million of them.

The Model T Revolution

The Model T was more than a mechanical marvel; it was a symbol of accessibility and freedom. Before its advent, automobiles were luxury items, affordable only to the wealthy. The Model T changed that paradigm. Priced initially at $850, and eventually dropping to as low as $300, it put car ownership within reach of the average American.

Thomas Bennett, a young and ambitious engineer, was among those who saw the Model T not merely as a machine but as a transformative force. Born in 1890, Thomas grew up in a small town in Ohio, where his fascination with mechanics was evident from a young age. He spent countless hours tinkering with anything that had moving parts, from pocket watches to steam engines. His passion led him to pursue a degree in mechanical engineering, and upon graduation, he found himself irresistibly drawn to Detroit, the epicenter of the automotive revolution.

Thomas Bennett's Early Years

Detroit in the early 1900s was a city alive with energy and innovation. The air was thick with the smell of oil and the sound of machinery. It was here that Thomas Bennett began his career, joining Ford's burgeoning empire as a junior engineer. The experience was transformative. Thomas was captivated by the Model T's simplicity and robustness. He marveled at the assembly line, where each worker played a precise role in the creation of a finished product.

Thomas's talents did not go unnoticed. His keen eye for detail and relentless work ethic quickly earned him a reputation as a rising star. Within a few years, he advanced to a supervisory position, overseeing a team dedicated to refining the production process. His innovations in tool design and workflow optimization contributed to even greater efficiencies on the assembly line.

Building Dreams

Despite his success at Ford, Thomas harbored dreams of something more. He envisioned a vehicle that combined the affordability of the Model T with enhanced performance and aesthetics. In 1921, with a modest savings and an unwavering determination, he left Ford to start his own venture. The Bennett Workshop was born in a rented garage on the outskirts of Detroit.

The early days were challenging. Thomas's resources were limited, and he often worked long hours, fueled by little more than black coffee and sheer willpower. His wife, Margaret, managed the finances and kept the household running smoothly, providing a stable foundation for his endeavors. Together, they formed a formidable team.

Thomas's first creation, the Bennett Roadster, was a modest success. It was a sleek, two-seater that captured the spirit of the Roaring Twenties. Priced competitively, it appealed to young professionals and adventurers alike. The Roadster's success bolstered Thomas's reputation and allowed him to expand his operations.

Family Life and the Bennett Workshop

Thomas and Margaret's partnership extended beyond the business. Their home was a sanctuary of support and shared dreams. Margaret's unwavering belief in Thomas's vision was the bedrock of their relationship. She managed the household with grace, ensuring that their children, Emily and Jack, felt secure and loved despite the long hours Thomas spent in the workshop.

Emily, the elder of the two, was a curious and imaginative child. She often sat with her father in the garage, asking endless questions about the cars he built. Her interest in design and mechanics was evident even at a young age. Jack, on the other hand, was a bundle of energy and enthusiasm. He loved anything that moved fast and had a knack for understanding how things worked.

On the Open Road

The automobile was changing America in profound ways. Roads that had once been dusty trails now transformed into paved highways, connecting cities and towns like never before. The concept of the road trip emerged, capturing the imagination of a nation eager to explore its vast landscapes.

For the Bennett family, road trips became a cherished tradition. Each summer, Thomas, Margaret, Emily, and Jack would pile into the latest Bennett model and set off on a journey of discovery. These trips were more than vacations; they were opportunities to test the limits of Thomas's designs and to bond as a family.

The Bennetts traversed the country, from the bustling streets of New York City to the serene beauty of the Grand Canyon. Along the way, they encountered a diverse array of people, each with their own stories and aspirations. These journeys fostered a deep appreciation for the transformative power of the automobile, not just as a means of transportation but as a catalyst for human connection and adventure.

The First Cross-Country Road Trip

In the summer of 1923, the Bennetts embarked on

their most ambitious road trip yet: a cross-country journey from Detroit to California. This trip was a celebration of the resilience they had shown through the Great Depression and a testament to the quality of Bennett automobiles. Thomas saw it as an opportunity to test his latest model, the Bennett Cruiser, under real-world conditions.

The Bennett Cruiser was a marvel of engineering, featuring a more powerful engine, improved suspension, and a host of comfort features designed for long-distance travel. Thomas was confident that this car would not only survive the journey but thrive, demonstrating the durability and reliability of his designs.

The family—Thomas, Margaret, Emily, and Jack—packed their bags and set off early one June morning. The car was loaded with provisions for the long journey: maps, a portable stove, camping gear, and plenty of food and water. Thomas and Margaret took turns driving, while Emily, now in her late teens, navigated using a set of maps spread across her lap. Jack, ever the adventurer, kept a journal of their travels, documenting every sight and experience.

Family Bonding and Roadside Adventures

The open road offered a unique opportunity for the Bennetts to bond as a family. They drove through bustling cities and quiet towns, across vast plains and over towering mountains. Each stop brought new adventures and discoveries. In Chicago, they marveled at the skyline and visited the bustling markets. In Kansas, they camped under the stars, telling stories around the campfire.

One of the most memorable stops was in Colorado, where they visited the Rocky Mountains. The air was crisp and clear, and the views were breathtaking. The family spent several days hiking and exploring, with Thomas taking the opportunity to test the Cruiser's performance on the winding mountain roads. The car handled the steep inclines and sharp turns with ease, reaffirming Thomas's confidence in his engineering.

As they continued westward, the landscape changed dramatically. The lush greenery of the Midwest gave way to the arid deserts of the Southwest. The family drove through vast stretches of desert, stopping occasionally to marvel at the stark beauty of the landscape. In Arizona, they visited the Grand Canyon, standing in awe at the sheer scale and majesty of the natural wonder.

Emily's Emerging Passion for Design

During the trip, Emily's interest in automotive design blossomed. She had always been curious about her father's work, often asking questions about the mechanics and engineering behind his creations. This journey provided her with a deeper appreciation for the practical challenges and the artistry involved in designing a car.

One evening, as they camped near a quiet lake in Utah, Emily and Thomas sat by the water, discussing the intricacies of car design. Thomas explained the principles of aerodynamics, the importance of balance and weight distribution, and the challenges of creating a vehicle that was both functional and beautiful.

Emily listened intently, her mind racing with ideas. She began to sketch her own designs, envisioning cars that combined elegance with performance. Thomas was impressed by her creativity and encouraged her to pursue her passion. He promised to mentor her and teach her everything he knew about automotive design.

Jack's Racing Dreams

Jack, on the other hand, was captivated by speed and adventure. The long stretches of open road fueled his imagination, and he dreamed of becoming a race car driver. He admired the precision and skill required to navigate a car at high speeds and was determined to one day compete in the prestigious races he had read about.

Throughout the trip, Jack peppered his father with questions about racing techniques, engine performance, and the physics of speed. Thomas, recognizing his son's enthusiasm, shared stories of legendary races and racers, providing Jack with a wealth of knowledge and inspiration.

One afternoon, while driving through the deserts of Nevada, Thomas allowed Jack to take the wheel on a long, straight stretch of highway. Jack's excitement was palpable as he accelerated, feeling the power of the Cruiser beneath him. Thomas guided him, teaching him the importance of control and precision, emphasizing that speed was not just about going fast but about mastering the vehicle.

A New Era Begins

The Bennett family's cross-country journey was more than just a road trip; it was a testament to the enduring spirit of exploration and innovation. The open road symbolized the limitless possibilities of the future, a future that the Bennetts were determined to shape.

As they settled back into their lives in Detroit, Thomas, Margaret, Emily, and Jack carried with them the memories and lessons of their adventure. The road ahead was filled with promise, and the Bennett family was ready to embrace it, continuing their legacy of excellence and innovation in the automotive world.

"The Road Taken" was more than a journey; it was a celebration of the enduring bond between people and their machines, a bond that would continue to drive the Bennetts forward on the open road.

Chapter 2: Building Dreams

Innovations in Detroit

The 1920s in Detroit were characterized by rapid industrial growth and a sense of limitless potential. The city, often referred to as the "Motor City," became the epicenter of the American automotive industry. Factories hummed with activity, producing cars that would soon populate the burgeoning network of roads across the nation. Amidst this flurry of innovation, Thomas Bennett's journey as an independent automotive entrepreneur began to take shape.

After leaving Ford to pursue his vision, Thomas set up the Bennett Workshop in a modest garage on the outskirts of Detroit. The early days were grueling. Thomas spent countless hours drafting blueprints, experimenting with materials, and assembling prototypes. He was a perfectionist, driven by a relentless pursuit of excellence. Every nut, bolt, and rivet had to meet his exacting standards.

Margaret Bennett played an indispensable role during these formative years. While Thomas focused on the technical aspects, Margaret managed the finances, handled customer inquiries, and provided unwavering emotional support. Her belief in Thomas's vision never wavered, even during the most challenging times.

The Bennett Workshop

The first significant breakthrough came in 1921 with the introduction of the Bennett Roadster. This sleek, two-seater car was designed for both style and performance. Thomas aimed to create a vehicle that was not only affordable but also visually appealing and enjoyable to drive. The Roadster's design featured a streamlined body, an innovative suspension system, and a powerful yet efficient engine.

Thomas's meticulous attention to detail paid off. The Bennett Roadster quickly gained a reputation for its reliability and performance. Word of mouth spread, and soon the small workshop was inundated with orders. The success of the Roadster allowed Thomas to expand his operations, moving from the cramped garage to a larger facility that could accommodate increased production.

Early Challenges and Triumphs

Despite the initial success, the path was not without obstacles. The automotive industry was fiercely competitive, and established companies like Ford, General Motors, and Chrysler dominated the market. As a small, independent manufacturer, Bennett Motors had to carve out its niche. Thomas's approach was to focus on quality and innovation, offering features that set his cars apart from the mass-produced models of the larger companies.

One such innovation was the development of a more

advanced braking system. In the early days of automotive manufacturing, brakes were often unreliable, leading to frequent accidents. Thomas devoted considerable time and resources to designing a braking system that was both effective and durable. The result was a safer driving experience, which became a significant selling point for Bennett cars.

Another challenge was financing. Expanding production required capital, and traditional banks were often reluctant to lend to small, unproven businesses. Thomas and Margaret faced several rejections before finally securing a loan from a local bank impressed by their perseverance and vision. This financial boost enabled them to purchase new machinery and hire additional workers, further accelerating their growth.

The Rise of Luxury Cars

As the Roaring Twenties progressed, the demand

for luxury and performance cars surged. The economic prosperity of the era led to a consumer culture that prized status symbols, and automobiles were at the forefront. Thomas recognized this trend and began designing cars that catered to the affluent market while maintaining the core values of quality and innovation that defined the Bennett brand.

In 1925, Bennett Motors introduced the Bennett Royale, a luxury sedan that epitomized elegance and sophistication. The Royale featured plush interiors, advanced suspension for a smooth ride, and a powerful engine that delivered both speed and reliability. It quickly became a favorite among the wealthy, including several prominent figures in business and entertainment.

The success of the Bennett Royale cemented Thomas's reputation as a visionary in the automotive industry. It also allowed the Bennett Workshop to expand further, with new facilities and a growing workforce dedicated to producing some of the finest automobiles of the era.

Thomas Bennett's Breakthrough Designs

Throughout the late 1920s, Thomas continued to innovate. One of his most significant contributions was the introduction of the Bennett Super Six engine. This engine combined advanced engineering with practical design, offering unparalleled performance and efficiency. It became the hallmark of Bennett cars, known for their powerful yet smooth driving experience.

Thomas also pioneered the use of lightweight materials in car construction. By incorporating aluminum and other alloys, he was able to reduce the weight of his vehicles, improving fuel efficiency and handling. These innovations kept Bennett Motors at the cutting edge of automotive technology, attracting a loyal customer base that appreciated the blend of style and substance.

The Golden Age of Automobiles

The 1920s were a golden age for automobiles, with cars becoming symbols of modernity and progress. The Bennett Royale and other luxury models captured the imagination of the public, appearing in advertisements, films, and popular culture. Thomas Bennett's creations were more than just vehicles; they were statements of elegance and technological prowess.

The Bennett family enjoyed the fruits of their labor, but they never lost sight of their humble beginnings. They remained committed to their core values of quality, innovation, and customer satisfaction. Thomas and Margaret instilled these values in their children, Emily and Jack, who were beginning to take a more active interest in the family business.

Emily and Jack's Growing Roles

Emily, now in her early twenties, was deeply

involved in the design process. Her sketches and ideas were increasingly influencing the look and feel of Bennett cars. She had a keen eye for aesthetics and a talent for integrating artistic elements with practical design. Thomas encouraged her creativity, recognizing that her contributions would be vital to the company's future.

Jack, on the other hand, was captivated by the mechanical aspects of car manufacturing. He spent countless hours in the workshop, learning from the engineers and mechanics. His fascination with speed and performance led him to experiment with new engine designs and modifications, some of which were incorporated into Bennett's high-performance models.

The Impact of the Great Depression

As the decade drew to a close, the economic landscape began to shift. The stock market crash of 1929 ushered in the Great Depression, bringing widespread financial hardship. The automotive industry, like many others, faced significant challenges. Sales plummeted, and many companies struggled to stay afloat.

For Bennett Motors, the impact was profound. Orders declined sharply, and the company faced mounting financial pressure. Thomas and Margaret had to make difficult decisions to keep the business running. They reduced production, laid off workers, and sought new ways to cut costs without compromising quality.

Surviving the Storm

The Great Depression tested the resilience of the Bennett family. Thomas worked tirelessly to adapt to the changing market, exploring opportunities to diversify his products and find new revenue streams. Margaret continued to manage the finances, ensuring that every penny was wisely spent.

Despite the challenges, the Bennetts remained committed to their vision. They introduced a more affordable line of cars, designed to meet the needs of a struggling population. These models, while less luxurious than the Royale, still embodied the craftsmanship and reliability that defined the Bennett brand.

During these difficult years, Thomas's ingenuity shone through. He focused on developing technologies that improved fuel efficiency and reduced maintenance costs, addressing the concerns of cost-conscious consumers. These innovations helped Bennett Motors survive the economic downturn and positioned the company for recovery as the economy began to stabilize.

The Bennett Family's Struggles

The Great Depression also took a personal toll on the Bennett family. The stress of maintaining the business and supporting their employees weighed heavily on Thomas and Margaret. They worked long hours and made countless sacrifices to keep Bennett Motors afloat. Their determination and perseverance were a testament to their unwavering commitment to their family and their employees.

Emily and Jack, now young adults, took on more responsibilities within the company. Emily continued to contribute to the design process, while Jack focused on improving production efficiency and exploring new technologies. Their efforts were crucial in helping the company navigate these challenging times.

Looking to the Future

By the mid-1930s, the worst of the Depression had passed, and Bennett Motors emerged stronger and more resilient. The company's ability to innovate and adapt had been tested, and it had proven its mettle. Thomas Bennett's unwavering commitment to quality and innovation, supported by Margaret's financial acumen, ensured the survival and continued success of their enterprise.

As Thomas looked ahead, he saw new opportunities on the horizon. The automotive industry was poised for another wave of transformation, with advances in technology and changing consumer preferences driving new trends. For Bennett Motors, the road ahead was filled with promise.

The Bennett family's journey was far from over. They had weathered the storm of economic turmoil and emerged with a renewed sense of purpose. The dreams that had driven Thomas to leave Ford and start his own company were alive and well, fueled by a passion for excellence and a belief in the power of innovation.

With a firm foundation and a clear vision, Bennett Motors was ready to embrace the future, continuing its legacy of building dreams on the open road.

Chapter 3: On the Open Road

America's Love Affair with Cars

As the 1930s gave way to the 1940s, America's relationship with the automobile deepened. The car was no longer just a mode of transportation; it had become an integral part of American identity and culture. The open road symbolized freedom, adventure, and the promise of new horizons. Families took to the highways, exploring the vast and varied landscapes of the United States. For the Bennett family, these road trips were not just vacations but essential experiences that shaped their lives and the development of Bennett Motors.

The First Cross-Country Road Trip

In the summer of 1940, the Bennetts embarked on their most ambitious road trip yet: a cross-country journey from Detroit to California. This trip was a celebration of the resilience they had shown through the Great Depression and a testament to the quality of Bennett automobiles. Thomas saw it as an opportunity to test his latest model, the Bennett Cruiser, under real-world conditions.

The Bennett Cruiser was a marvel of engineering, featuring a more powerful engine, improved suspension, and a host of comfort features designed for long-distance travel. Thomas was confident that this car would not only survive the journey but thrive, demonstrating the durability and reliability of his designs.

The family—Thomas, Margaret, Emily, and Jack—packed their bags and set off early one June morning. The car was loaded with provisions for the long journey: maps, a portable stove, camping gear, and plenty of food and water. Thomas and Margaret took turns driving, while Emily, now in her late teens, navigated using a set of maps spread across her lap. Jack, ever the adventurer, kept a journal of their travels, documenting every sight and experience.

Family Bonding and Roadside Adventures

The open road offered a unique opportunity for the Bennetts to bond as a family. They drove through bustling cities and quiet towns, across vast plains and over towering mountains. Each stop brought new adventures and discoveries. In Chicago, they marveled at the skyline and visited the bustling markets. Emily found inspiration in the city's architecture, sketching ideas for future car designs. In Kansas, they camped under the stars, telling stories around the campfire and roasting marshmallows. Jack's journal entries from this time were filled with excitement and wonder.

One of the most memorable stops was in Colorado, where they visited the Rocky Mountains. The air was crisp and clear, and the views were breathtaking. The family spent several days hiking and exploring, with Thomas taking the opportunity to test the

Cruiser's performance on the winding mountain roads. The car handled the steep inclines and sharp turns with ease, reaffirming Thomas's confidence in his engineering.

As they continued westward, the landscape changed dramatically. The lush greenery of the Midwest gave way to the arid deserts of the Southwest. The family drove through vast stretches of desert, stopping occasionally to marvel at the stark beauty of the landscape. In Arizona, they visited the Grand Canyon, standing in awe at the sheer scale and majesty of the natural wonder. Emily's sketches captured the dramatic vistas, while Jack's journal entries reflected his awe and excitement.

Emily's Emerging Passion for Design

During the trip, Emily's interest in automotive design blossomed. She had always been curious about her father's work, often asking questions about the mechanics and engineering behind his creations. This journey provided her with a deeper appreciation for the practical challenges and the artistry involved in designing a car.

One evening, as they camped near a quiet lake in Utah, Emily and Thomas sat by the water, discussing the intricacies of car design. Thomas explained the principles of aerodynamics, the importance of balance and weight distribution, and the challenges of creating a vehicle that was both functional and beautiful.

"I've been thinking about ways to make cars more aerodynamic," Emily said, sketching in her notebook. "What if we designed the body to flow more naturally, like the curves of this lake?"

Thomas looked at her sketch and nodded, impressed by her creativity. "That's an excellent idea, Emily. The key is to reduce drag while maintaining stability. We can experiment with different shapes and materials when we get back."

Emily listened intently, her mind racing with ideas. She began to sketch her own designs, envisioning cars that combined elegance with performance. Thomas was impressed by her creativity and encouraged her to pursue her passion. He promised to mentor her and teach her everything he knew about automotive design.

Jack's Racing Dreams

Jack, on the other hand, was captivated by speed and adventure. The long stretches of open road fueled his imagination, and he dreamed of becoming a race car driver. He admired the precision and skill required to navigate a car at high speeds and was determined to one day compete in the prestigious races he had read about.

Throughout the trip, Jack peppered his father with questions about racing techniques, engine performance, and the physics of speed. Thomas, recognizing his son's enthusiasm, shared stories of legendary races and racers, providing Jack with a wealth of knowledge and inspiration.

One afternoon, while driving through the deserts of Nevada, Thomas allowed Jack to take the wheel on a long, straight stretch of highway. Jack's excitement was palpable as he accelerated, feeling the power of the Cruiser beneath him. Thomas guided him, teaching him the importance of control and precision, emphasizing that speed was not just about going fast but about mastering the vehicle.

"Remember, Jack, it's not just about how fast you can go, but how well you can handle the car," Thomas said as Jack navigated the desert road. "Racing is as much about skill and strategy as it is about speed."

Jack nodded, absorbing every word. This hands-on experience solidified his dream of becoming a race car driver and ignited a lifelong passion for high-performance vehicles.

The Pacific Coast

After weeks on the road, the Bennetts finally reached California. The sight of the Pacific Ocean was a triumphant moment, a testament to their perseverance and the reliability of the Bennett Cruiser. They spent several days exploring the coastal cities, from the bustling streets of Los Angeles to the scenic beauty of San Francisco.

In Los Angeles, they visited a prestigious automotive exhibition, where Thomas showcased the Bennett Cruiser. The car garnered significant attention, with many impressed by its performance and design. This exposure helped elevate the Bennett brand, attracting new customers and solidifying the company's reputation for quality and innovation.

Emily and Jack were thrilled by the bustling city and the vibrant car culture. Emily took note of the sleek, modern designs on display, while Jack marveled at the powerful engines and high-performance vehicles. The exhibition was a source of inspiration for both siblings, fueling their respective passions for design and racing.

Homeward Bound

The return journey to Detroit was equally memorable, filled with more adventures and discoveries. The family took a different route, exploring new states and cities. By the time they arrived home, they had traveled over 6,000 miles, a journey that had tested the limits of both the car and its occupants.

The cross-country trip was a defining moment for the Bennett family. It strengthened their bond and provided invaluable experiences that would shape their future. For Thomas, it was a validation of his life's work, proving that his designs could endure the rigors of the open road. For Margaret, it was a celebration of family and resilience, a reminder of the strength they had shown through challenging times.

For Emily and Jack, the journey was transformative. Emily returned with a renewed passion for design, eager to apply what she had learned from her father. Jack's dreams of racing were more vibrant than ever, fueled by the thrill of the road and the knowledge he had gained.

On the Open Road

The Bennett family's cross-country journey was more than just a road trip; it was a testament to the enduring spirit of exploration and innovation. The open road symbolized the limitless possibilities of the future, a future that the Bennetts were determined to shape.

As they settled back into their lives in Detroit, Thomas, Margaret, Emily, and Jack carried with them the memories and lessons of their adventure. The road ahead was filled with promise, and the Bennett family was ready to embrace it, continuing their legacy of excellence and innovation in the automotive world.

"The Road Taken" was more than a journey; it was a celebration of the enduring bond between people and their machines, a bond that would continue to drive the Bennetts forward on the open road.

Chapter 4: The Golden Age of Automobiles

The Rise of Luxury Cars

As the world recovered from the Great Depression and emerged from the turmoil of World War II, the 1950s ushered in a period of unprecedented economic prosperity and cultural change in the United States. This era, often referred to as the Golden Age of Automobiles, saw the rise of luxury cars that symbolized status, wealth, and the American dream. For Bennett Motors, this period was marked by significant innovation and expansion.

The Bennett family had weathered the economic challenges of the previous decades with resilience and ingenuity. Thomas Bennett's dedication to quality and innovation had established a strong foundation for the company. Now, with the economy booming and consumer demand for luxury at an all-time high, Bennett Motors was poised to take the next step.

Thomas Bennett's Breakthrough Designs

Thomas Bennett continued to push the boundaries of automotive design. He understood that the post-war consumer was looking for more than just a reliable vehicle; they wanted a car that embodied style, comfort, and modernity. With this in mind, he set out to create a line of luxury cars that would captivate the market.

In 1952, Bennett Motors introduced the Bennett Imperial, a flagship model that epitomized elegance and sophistication. The Imperial featured sleek, aerodynamic lines, a powerful V8 engine, and a luxurious interior with the latest amenities. Every detail, from the hand-stitched leather seats to the state-of-the-art sound system, was designed to provide an unparalleled driving experience.

Thomas's meticulous attention to detail paid off. The

Bennett Imperial quickly gained a reputation for its reliability and performance. Word of mouth spread, and soon the company was inundated with orders. The success of the Imperial allowed Bennett Motors to expand its operations, opening new manufacturing plants and showrooms across the country.

Emily Bennett's Influence on Car Design

Emily Bennett, now in her twenties, had taken a more active role in the family business. Her passion for design, nurtured during the family's cross-country road trips, had blossomed into a career. Emily brought a fresh perspective to Bennett Motors, combining her father's technical expertise with her own creative flair.

Emily's first major project was the Bennett Elegance, a mid-sized luxury sedan introduced in 1955. The Elegance was designed to appeal to a broader audience, offering a more affordable option without compromising on quality or style. Emily's influence was evident in the car's bold color choices, innovative use of materials, and attention to ergonomic design.

The Elegance was a hit, particularly among young professionals and families. It received rave reviews for its blend of style, performance, and affordability. Emily's contributions helped solidify her reputation as a talented designer and marked the beginning of a new era for Bennett Motors.

Women in the Automotive World

Emily's success also highlighted the changing role of women in the automotive industry. Traditionally a male-dominated field, the industry was beginning to recognize the value of diverse perspectives. Emily became a trailblazer, inspiring other women to pursue careers in automotive design and engineering.

She frequently spoke at industry conferences and participated in design exhibitions, sharing her experiences and advocating for greater inclusion of women in the field. Her efforts helped pave the way for a more inclusive and innovative industry, where talent and creativity were recognized regardless of gender.

The 1950s and 1960s Car Culture

The 1950s and 1960s were a time of cultural transformation in America, and cars played a central role in this shift. The rise of car culture was reflected in music, movies, and the burgeoning popularity of drive-in theaters and diners. Cars became more than just a means of transportation; they were expressions of individuality and freedom.

Bennett Motors capitalized on this cultural shift by producing cars that resonated with the youth market. The Bennett Rebel, introduced in 1960, was a sporty coupe designed for young drivers seeking excitement and style. With its powerful engine, sleek design, and affordable price, the Rebel became an instant hit.

Jack Bennett, now a charismatic young man with a passion for racing, played a key role in promoting the Rebel. He participated in numerous racing events and car shows, demonstrating the Rebel's capabilities and building a loyal following. Jack's involvement not only boosted sales but also reinforced the Bennett brand's association with performance and excitement.

The Family Road Trip Tradition

Despite their busy schedules, the Bennett family continued their tradition of road trips. These journeys were more than just vacations; they were opportunities to test new models, gather feedback, and strengthen family bonds. The trips also provided inspiration for future designs, as Thomas, Emily, and Jack observed how their cars performed in real-world conditions.

One memorable trip took the family along the Pacific Coast Highway, from San Francisco to Los Angeles. The winding coastal roads, breathtaking views, and diverse landscapes offered the perfect backdrop for testing the latest Bennett models. The family camped on the beaches, explored coastal towns, and enjoyed the camaraderie that these trips always fostered.

For Emily, these trips were a source of endless inspiration. She carried a sketchbook with her, capturing ideas for new designs influenced by the natural beauty and unique character of each place they visited. Jack, ever the thrill-seeker, reveled in the driving challenges posed by the twisty roads and steep inclines, providing valuable feedback on the cars' performance.

Route 66 and the American Dream

The iconic Route 66, known as the "Main Street of America," played a significant role in the Bennett family's adventures. This historic highway, stretching from Chicago to Los Angeles, epitomized the spirit of the open road and the American dream. The Bennetts often traveled segments of Route 66, immersing themselves in the rich history and culture that the route represented.

They visited landmarks like the Cadillac Ranch in Texas, the Wigwam Motel in Arizona, and the Santa Monica Pier at the western terminus of the highway. These journeys along Route 66 were a testament to the enduring allure of the road trip and the deep connection between the automobile and American culture.

Social Changes Reflected in Cars

The 1960s were a decade of profound social change,

and these shifts were reflected in the automotive industry. The civil rights movement, the rise of counterculture, and growing environmental awareness all influenced consumer preferences and automotive design.

Bennett Motors responded to these changes by introducing more diverse and environmentally friendly models. The Bennett Eco, launched in 1968, was one of the first hybrid vehicles, combining a traditional gasoline engine with an electric motor. This innovation was driven by Thomas's forward-thinking vision and Emily's commitment to sustainability.

The Eco was designed to appeal to environmentally conscious consumers, offering a blend of efficiency and performance. It received acclaim for its innovative technology and stylish design, positioning Bennett Motors as a leader in the emerging field of green automotive solutions.

Innovations and Market Expansion

Thomas and Emily continued to push the boundaries of automotive design and technology. They introduced several innovations that set Bennett cars apart from the competition. These included advanced suspension systems for a smoother ride, enhanced safety features such as reinforced frames and improved braking systems, and more efficient engines that delivered better performance and fuel economy.

Bennett Motors also expanded its market reach, opening new dealerships across the United States and exploring opportunities for international sales. The company's reputation for quality and innovation attracted a loyal customer base, and sales soared. By the early 1970s, Bennett Motors had firmly established itself as a leader in the automotive industry.

Jack Bennett's Return and Racing Dreams

After serving in the military, Jack Bennett returned home with a newfound sense of purpose and a passion for racing. Inspired by his experiences and the technological advancements he had witnessed during the war, Jack was determined to make his mark in the world of motorsports. He joined Bennett Motors, bringing with him a wealth of knowledge and a competitive spirit.

Jack's influence was instrumental in the development of the Bennett Racer, a high-performance sports car designed for the track. The Racer combined cutting-edge technology with sleek, aerodynamic design, making it a formidable contender in the racing world. Jack competed in numerous races, earning accolades and bringing further prestige to the Bennett brand.

The Impact of the Interstate Highway System

The post-war era also saw the construction of the Interstate Highway System, a network of roads that transformed transportation in the United States. The new highways made long-distance travel easier and more accessible, fueling the growth of car culture and increasing demand for reliable and comfortable vehicles.

Bennett Motors recognized the opportunities presented by the Interstate Highway System and developed cars designed for long-distance travel. These vehicles featured larger fuel tanks, improved suspension systems, and enhanced comfort features, making them ideal for road trips and cross-country journeys.

The Bennett family continued their tradition of road trips, exploring the new highways and testing their latest models on long journeys. These trips provided valuable insights into the performance and durability of Bennett cars, helping the company refine and improve their designs.

As the 1960s came to a close, the Bennetts looked to the future with optimism and determination. The road ahead promised new challenges and opportunities, and they were ready to embrace them, continuing their legacy of excellence and innovation in the world of automobiles.

"The Golden Age of Automobiles" marked a period of transformation and growth for Bennett Motors. The company's commitment to quality, style, and performance resonated with consumers, establishing Bennett as a premier brand in the automotive industry. The Bennetts' dedication to their craft and their unwavering pursuit of excellence ensured that Bennett Motors would continue to thrive, leading the way into a new era of automotive innovation.

Chapter 5: Surviving the Great Depression

Economic Turmoil and the Auto Industry

The roaring twenties had brought prosperity and growth, but the dawn of the 1930s painted a starkly different picture. The Great Depression cast a long shadow over the nation, plunging millions into unemployment and despair. For the automotive industry, which had been riding high on the wave of consumerism, the impact was devastating. Car sales plummeted, factories closed, and many companies faced bankruptcy.

For Bennett Motors, the effects were immediate and severe. Orders for new cars dried up almost overnight, leaving the company with excess inventory and mounting financial pressures. Thomas Bennett, ever the optimist and problem-solver, knew that he had to take drastic measures to ensure the survival of his company.

The Bennett Family's Struggles

The Bennett family faced the harsh realities of the Great Depression with a mixture of resilience and determination. Thomas and Margaret Bennett had always been prudent with their finances, but even their careful planning could not shield them entirely from the economic storm. The family tightened their belts, cutting back on non-essential expenses and finding ways to make do with less.

Thomas was forced to lay off a significant portion of his workforce, a decision that weighed heavily on him. He had always prided himself on providing good jobs and fostering a sense of community within his company. Watching skilled workers and loyal employees lose their livelihoods was one of the hardest experiences of his career.

Despite these challenges, Thomas remained committed to his vision of building quality automobiles. He spent long hours in the workshop, seeking ways to innovate and improve efficiency. Margaret managed the household and the company's finances, ensuring that every dollar was stretched as far as possible. Their children, Emily and Jack, pitched in where they could, understanding the gravity of the situation and the importance of family solidarity.

Innovations in Tough Times

In the face of adversity, Thomas Bennett's ingenuity and creativity shone through. He knew that to survive the Depression, Bennett Motors needed to adapt and find new ways to appeal to cash-strapped consumers. One of his first steps was to introduce a more affordable line of cars that retained the quality and reliability the Bennett brand was known for.

In 1931, Bennett Motors launched the Bennett Compact, a smaller, more economical car designed to meet the needs of a struggling population. The Compact featured a simplified design, with fewer luxury features but maintained the durability and performance that were hallmarks of Bennett vehicles. Priced competitively, the Compact appealed to families and individuals who needed reliable transportation at a lower cost.

The introduction of the Bennett Compact was a strategic move that helped stabilize the company's finances. Sales of the Compact provided a much-needed influx of cash, allowing Bennett Motors to keep its doors open and retain a core group of skilled workers. Thomas also focused on improving the manufacturing process, finding ways to reduce waste and increase productivity without compromising quality.

Personal Sacrifices and Community Support

The Great Depression was a time of immense hardship for many families, and the Bennetts were no exception. Thomas and Margaret made personal sacrifices to ensure the survival of their company and the well-being of their employees. They sold their second car, cut back on household expenses, and dipped into their savings to keep the business afloat.

Margaret's role became even more crucial during this period. She took on additional responsibilities, managing not only the household but also the company's finances. Her meticulous budgeting and financial acumen were vital in navigating the economic downturn. Margaret's strength and resilience provided a stable foundation for the family during these challenging times.

The Bennett family also found support in their local community. Neighbors and friends rallied around them, offering help and encouragement. The community's faith in Bennett Motors and the quality of its vehicles never wavered. This support bolstered the family's spirits and reinforced their determination to overcome the hardships they faced.

Innovations and Adaptations

Thomas Bennett's relentless pursuit of innovation did not falter during the Depression. He continued to explore new technologies and materials that could improve the efficiency and affordability of his cars. One of his significant innovations was the development of a more fuel-efficient engine, which became a key selling point for the Bennett Compact.

Thomas also experimented with alternative materials to reduce production costs. He sought out suppliers who could provide high-quality materials at lower prices and implemented cost-saving measures throughout the manufacturing process. These efforts helped Bennett Motors maintain its reputation for quality while adapting to the economic constraints of the time.

Emily and Jack played active roles in these innovations. Emily's design skills were instrumental in creating the streamlined, simplified aesthetic of the Bennett Compact. She worked closely with her father to ensure that the car's design was both attractive and functional. Jack, on the other hand, focused on improving production efficiency. He introduced new methods and techniques that reduced waste and increased productivity, helping the company weather the economic storm.

Community Outreach and Corporate Responsibility

Despite their own struggles, the Bennett family was committed to giving back to their community. Thomas and Margaret believed in the importance of corporate responsibility and sought ways to support those who were also suffering during the Depression. Bennett Motors launched several initiatives aimed at helping the local community.

One such initiative was a job training program for unemployed workers. Thomas and his team provided training in automotive repair and maintenance, giving people the skills they needed to find work. The program was a success, helping many individuals regain their sense of purpose and stability.

Margaret organized community events and fundraisers to support local charities and relief efforts. These events not only provided much-needed aid but also brought the community closer together. The Bennetts' commitment to their community earned them respect and admiration, further solidifying their reputation as leaders both in business and in society.

The Road to Recovery

As the 1930s drew to a close, the worst of the Depression began to lift, and the American economy started to recover. Bennett Motors emerged from the crisis stronger and more resilient. The company's ability to innovate and adapt had been tested, and it had proven its mettle. Thomas

Bennett's unwavering commitment to quality and innovation, supported by Margaret's financial acumen, ensured the survival and continued success of their enterprise.

The success of the Bennett Compact during the Depression laid the groundwork for future growth. The company had built a loyal customer base that appreciated the quality and reliability of Bennett vehicles. As the economy improved, Bennett Motors was well-positioned to expand its product line and capitalize on new opportunities.

Emily and Jack had grown into their roles within the company, bringing fresh perspectives and ideas. Emily's innovative designs and Jack's technical expertise were invaluable assets, driving the company's continued success. The Bennett family's collective efforts ensured that Bennett Motors remained a leader in the automotive industry.

Looking to the Future

With the economy on the upswing, Bennett Motors was ready to embrace the future. Thomas Bennett saw new opportunities on the horizon, driven by advances in technology and changing consumer preferences. The company began developing new models that incorporated the latest innovations in automotive engineering.

The introduction of the Bennett Victory in 1940 marked a significant milestone. This new model featured a sleek design, advanced safety features, and improved fuel efficiency. The Victory was well-received by consumers and critics alike, solidifying Bennett Motors' reputation for quality and innovation.

As Thomas looked ahead, he saw a bright future for Bennett Motors. The experiences and lessons learned during the Great Depression had strengthened the company and prepared it for the challenges and opportunities that lay ahead. The Bennett family's journey was far from over, and they were ready to continue their legacy of excellence and innovation in the automotive world.

Chapter 6: World War II and the Automotive Shift

The War Effort and Automotive Production

As the 1930s came to an end, the world was gripped by escalating tensions that would soon lead to World War II. The United States, initially maintaining a stance of neutrality, eventually found itself drawn into the conflict following the attack on Pearl Harbor in December 1941. The war had a profound impact on every aspect of American life, including the automotive industry.

Bennett Motors, like many other manufacturers, was called upon to support the war effort. The company's production facilities were retooled to produce military vehicles, aircraft components, and other essential supplies. This transition required a significant shift in focus and resources, but Thomas Bennett was committed to doing his part for the country.

Thomas and Emily Bennett worked tirelessly to ensure that Bennett Motors could meet the demands of wartime production. The company's engineers and workers adapted quickly to the new requirements, producing everything from trucks and jeeps to airplane parts. The skills and innovations developed during this period would later prove invaluable in the post-war automotive boom.

Emily Bennett's Role in Wartime Production

With many men, including her brother Jack, serving in the military, Emily Bennett took on a more prominent role at Bennett Motors. Her background in design and engineering, combined with her experience working alongside her father, made her well-suited to handle the challenges of wartime production.

Emily's responsibilities included overseeing the production lines, ensuring quality control, and implementing new manufacturing techniques to improve efficiency. She also played a key role in designing military vehicles, applying her creative and technical skills to develop solutions that met the rigorous demands of combat conditions.

One of Emily's notable contributions was the design of a lightweight, all-terrain vehicle that became a staple of the U.S. military. Her innovative approach to materials and construction resulted in a vehicle that was both durable and easy to maneuver, earning praise from soldiers and military officials alike.

Jack Bennett's Military Service

While Emily managed the home front, Jack Bennett served with distinction in the military. His experiences during the war had a profound impact on him, shaping his outlook and deepening his appreciation for the role of technology and engineering in modern warfare.

Jack was assigned to a unit responsible for maintaining and repairing military vehicles. This hands-on experience gave him a deeper understanding of the practical challenges faced by soldiers in the field and the importance of reliability and durability in vehicle design. He also had the opportunity to witness firsthand the technological advancements being made in aviation and other areas of military engineering.

These experiences fueled Jack's passion for innovation and performance, setting the stage for his future contributions to Bennett Motors. He returned home with a wealth of knowledge and a renewed sense of purpose, eager to apply what he had learned to the family business.

Post-War Transformation and Expansion

With the end of World War II in 1945, the world entered a period of reconstruction and economic growth. The United States, having emerged from the war as a global superpower, experienced an economic boom that transformed every aspect of society. For the automotive industry, this period marked the beginning of a new era of innovation and expansion.

Bennett Motors, having successfully navigated the challenges of wartime production, was well-positioned to capitalize on the post-war boom. The company quickly transitioned back to producing consumer vehicles, leveraging the skills and technologies developed during the war to create more advanced and desirable cars.

In 1947, Bennett Motors introduced the Bennett Victory, a sleek and modern car that symbolized the optimism and progress of the post-war era. The

Victory featured a streamlined design, a powerful yet efficient engine, and a range of new amenities that appealed to a burgeoning middle class eager to embrace the future.

The Bennett Victory and Consumer Demand

The Bennett Victory was an immediate success, capturing the imagination of the American public. Its stylish design and advanced features made it a sought-after vehicle, and Bennett Motors struggled to keep up with demand. The car's success helped drive the company's expansion, allowing Bennett Motors to open new manufacturing plants and dealerships across the country.

The Victory's popularity was fueled by a combination of factors. The post-war economic boom meant that more Americans had disposable income and were eager to spend it on consumer goods, including automobiles. Additionally, the rise of suburbia and the increasing importance of car ownership in American life created a growing market for reliable and stylish vehicles.

Thomas Bennett and his team continued to innovate, introducing new models and features that kept Bennett Motors at the forefront of the industry. They focused on improving performance, safety, and comfort, responding to the evolving needs and desires of consumers.

The Role of Marketing and Advertising

The post-war era also saw the rise of modern marketing and advertising techniques, and Bennett Motors was quick to adopt these strategies. The company launched a series of high-profile advertising campaigns that showcased the elegance and sophistication of Bennett cars. These campaigns, featuring print ads, radio spots, and eventually television commercials, helped build the Bennett brand and attract new customers.

One particularly memorable campaign featured a series of advertisements highlighting the Bennett Victory's advanced safety features. These ads emphasized the car's reinforced frame, improved braking system, and innovative safety belts, appealing to families who prioritized safety and reliability.

The company also sponsored popular radio and television programs, aligning the Bennett brand with the glamour and excitement of Hollywood. These efforts helped create a sense of prestige and desirability around Bennett cars, reinforcing their status as symbols of success and modernity.

Emily Bennett's Continued Influence

Emily Bennett continued to play a crucial role in the company's success. Her innovative designs and forward-thinking approach helped shape the direction of Bennett Motors. In addition to her work on consumer vehicles, Emily was deeply involved in the development of new technologies and features that set Bennett cars apart from the competition.

One of her notable achievements was the introduction of the Bennett Deluxe, a luxury sedan that combined elegance with cutting-edge technology. The Deluxe featured an advanced suspension system, a powerful V8 engine, and a host of luxury amenities, including a state-of-the-art sound system and climate control.

Emily's designs were not only aesthetically pleasing but also highly functional. She was a pioneer in ergonomic design, focusing on creating comfortable and intuitive interiors that enhanced the driving experience. Her work earned her numerous accolades and cemented her reputation as one of the leading designers in the automotive industry.

Jack Bennett's Racing Career

After returning from the war, Jack Bennett pursued his passion for racing with renewed vigor. He joined the company's engineering team, where he applied his knowledge of high-performance vehicles to the development of the Bennett Racer, a sports car designed for the track.

The Bennett Racer was a testament to Jack's skill and determination. It featured a lightweight frame, a powerful engine, and aerodynamic design, making it a formidable competitor in the racing world. Jack's racing career quickly took off, and he became a well-known figure on the racing circuit.

Jack's success on the track brought further prestige to the Bennett brand. His victories in high-profile races showcased the performance and reliability of Bennett cars, attracting a new generation of enthusiasts. Jack also used his platform to advocate for safety and innovation, emphasizing the importance of quality engineering in both racing and consumer vehicles.

The Impact of the Interstate Highway System

The post-war era also saw the construction of the Interstate Highway System, a network of roads that transformed transportation in the United States. The new highways made long-distance travel easier and more accessible, fueling the growth of car culture and increasing demand for reliable and comfortable vehicles.

Bennett Motors recognized the opportunities presented by the Interstate Highway System and developed cars designed for long-distance travel. These vehicles featured larger fuel tanks, improved suspension systems, and enhanced comfort features, making them ideal for road trips and cross-country journeys.

The Bennett family continued their tradition of road

trips, exploring the new highways and testing their latest models on long journeys. These trips provided valuable insights into the performance and durability of Bennett cars, helping the company refine and improve their designs.

Emily Bennett's Entry into the Industry

During the war years, Emily Bennett's involvement in the family business deepened. With many men, including her brother Jack, serving in the armed forces, women stepped up to fill roles traditionally held by men. Emily, now in her twenties, took on more responsibilities at Bennett Motors, working alongside her father to oversee production and ensure quality control.

Emily's contributions were invaluable. Her keen eye for detail and innovative thinking helped streamline the production process and improve efficiency. She also played a role in designing military vehicles, applying her design skills to create functional and effective solutions for the war effort. Emily's work during this period earned her respect within the company and the broader automotive community.

Post-War Automotive Boom

With the end of World War II in 1945, the world entered a new era of peace and prosperity. The post-war economic boom brought renewed consumer demand for automobiles, and Bennett Motors was well-positioned to capitalize on this growth. The company quickly transitioned back to producing consumer vehicles, leveraging the innovations and efficiencies developed during the war.

In 1947, Bennett Motors introduced the Bennett Victory, a stylish and modern car that symbolized the optimism and progress of the post-war era. The Victory featured a sleek design, advanced safety features, and a powerful yet efficient engine. It was an immediate success, capturing the imagination of a public eager to embrace the future.

The Impact of World War II

The period following World War II was one of transformation and growth for Bennett Motors. The company's ability to adapt to the demands of wartime production and capitalize on the post-war economic boom ensured its continued success. Thomas Bennett's visionary leadership, Emily's innovative designs, and Jack's racing prowess all contributed to the company's enduring legacy.

As the world moved into the 1950s, Bennett Motors remained at the forefront of the automotive industry, embracing new technologies and meeting the evolving needs of consumers. The road ahead was filled with promise, and the Bennett family was ready to embrace it, continuing their journey of excellence and innovation in the world of automobiles.

Chapter 7: The Rise of Muscle Cars

The Birth of Iconic Models

The 1960s marked a significant shift in the automotive industry, characterized by the rise of muscle cars—high-performance vehicles known for their powerful engines and aggressive styling. This era of automotive history celebrated speed, power, and the thrill of driving. For Bennett Motors, this was a time of innovation and bold new designs.

Thomas Bennett, ever the visionary, recognized the growing demand for performance-oriented vehicles. He saw an opportunity to create cars that appealed to a younger, more adventurous demographic. Drawing on the company's rich heritage of engineering excellence, Bennett Motors set out to design a new line of muscle cars that would capture the spirit of the 1960s.

In 1965, Bennett Motors introduced the Bennett Fury,

a car that would become an icon of the muscle car era. The Fury featured a sleek, aerodynamic design, a powerful V8 engine, and a range of performance enhancements that made it a formidable contender on both the street and the track. Its aggressive styling, with bold lines and distinctive details, set it apart from other cars of the time.

Jack Bennett's Racing Career

Jack Bennett played a crucial role in the development and promotion of the Bennett Fury. His passion for racing and his extensive knowledge of high-performance engineering made him the perfect advocate for the new model. Jack's racing career had taken off in the post-war years, and he had become a well-known figure in the motorsports world.

Jack's involvement in the design process ensured that the Fury was more than just a powerful car; it was a finely tuned machine built for speed and performance. He tested the car extensively, providing valuable feedback on its handling, acceleration, and overall performance. His input helped refine the Fury, making it a benchmark for muscle cars.

To showcase the Fury's capabilities, Jack entered it in numerous racing events. The car's impressive performance on the track, coupled with Jack's racing prowess, garnered significant attention and acclaim. The Fury quickly became a favorite among enthusiasts, cementing Bennett Motors' reputation as a leader in the muscle car segment.

The Thrill of the Track

Racing had always been a part of Jack's life, but the 1960s saw him reach new heights in his career. He competed in a variety of events, from drag races to endurance races, each presenting its own set of challenges and thrills. The Bennett Fury was his trusted companion in these endeavors, and together they achieved numerous victories.

One of Jack's most memorable races took place at the Daytona International Speedway. The race was a grueling test of endurance and skill, with some of the best drivers and cars competing for the top spot. Jack's determination and the Fury's impeccable engineering proved to be a winning combination. After hours of intense competition, Jack crossed the finish line first, securing a major victory for Bennett Motors.

This win was not only a personal triumph for Jack but also a significant milestone for the company. The publicity generated by the victory at Daytona boosted sales of the Fury and solidified its status as an icon of the muscle car era. The Bennett brand became synonymous with speed, performance, and the excitement of the open road.

Emily Bennett's Influence on Car Design

While Jack focused on racing, Emily Bennett continued to drive innovation in car design. Her unique blend of artistic vision and technical expertise had already made a significant impact on the company's product lineup. During the 1960s, Emily turned her attention to the design of muscle cars, bringing a fresh perspective to this dynamic segment.

Emily's contributions to the Bennett Fury were substantial. She worked closely with the engineering team to ensure that the car's design was both functional and aesthetically pleasing. Her emphasis on aerodynamics, weight distribution, and ergonomics resulted in a car that was not only fast but also enjoyable to drive.

Emily also introduced a range of customization options for the Fury, allowing customers to personalize their vehicles. These options included a variety of paint colors, interior finishes, and performance upgrades. This focus on customization helped the Fury appeal to a broader audience, from hardcore racing enthusiasts to those seeking a stylish and powerful daily driver.

The Muscle Car Craze

The 1960s were a golden age for muscle cars, with manufacturers across the industry introducing models designed to capture the hearts of speed enthusiasts. The Bennett Fury stood out in this crowded market, thanks to its blend of performance, style, and innovation. It became a symbol of the era's rebellious spirit and the quest for freedom on the open road.

Car culture flourished during this time, with muscle cars playing a central role. The cars were celebrated in movies, music, and popular culture, becoming icons of American identity. Drive-in theaters, car shows, and drag races became popular pastimes, with muscle cars taking center stage.

For Bennett Motors, the success of the Fury and other muscle cars translated into significant growth. The company expanded its production facilities and dealership network to meet the rising demand. Sales soared, and Bennett Motors enjoyed a period of unprecedented prosperity.

The Environmental Movement and Industry Shifts

As the 1960s progressed into the early 1970s, the automotive industry began to face new challenges. The growing environmental movement raised concerns about air pollution, fuel consumption, and the impact of cars on the planet. These concerns led to increased regulatory scrutiny and the introduction of stricter emissions standards.

Bennett Motors, known for its innovation and adaptability, responded proactively to these changes. Thomas, Emily, and Jack recognized the need to balance performance with environmental responsibility. They began investing in research and development to create more fuel-efficient engines and cleaner technologies without sacrificing the excitement and power that defined their brand.

The introduction of the Bennett Eco in 1971 marked a significant milestone in this effort. The Eco was one of the first hybrid vehicles on the market, combining a traditional gasoline engine with an electric motor. This innovative approach significantly reduced emissions and improved fuel efficiency, appealing to environmentally conscious consumers.

The Bennett Legacy: Innovation and Excellence

The muscle car era left an indelible mark on the automotive industry and popular culture. The cars from this period, including the Bennett Fury, became timeless classics, celebrated for their design, performance, and the sense of freedom they embodied. Collectors and enthusiasts continue to cherish these vehicles, preserving them as symbols of a bygone era of automotive excellence.

For the Bennett family, the muscle car era was a time of growth, achievement, and adventure. The lessons learned and the innovations developed during this period laid the groundwork for the company's future success. The Bennett Fury and other models from this time remain enduring testaments to the family's passion for cars and their commitment to pushing the boundaries of what was possible.

The period following World War II was one of transformation and growth for Bennett Motors. The company's ability to adapt to the demands of wartime production and capitalize on the post-war economic boom ensured its continued success. Thomas Bennett's visionary leadership, Emily's innovative designs, and Jack's racing prowess all contributed to the company's enduring legacy.

As the 1950s progressed, Bennett Motors remained

at the forefront of the automotive industry, embracing new technologies and meeting the evolving needs of consumers. The road ahead was filled with promise, and the Bennett family was ready to embrace it, continuing their journey of excellence and innovation in the world of automobiles.

The story of Bennett Motors during the rise of muscle cars is a testament to the enduring appeal of speed, performance, and the open road. It celebrates the passion and creativity of the Bennett family and their unwavering dedication to building cars that excite and inspire. As they looked to the horizon, the Bennetts knew that their journey was far from over, and they were excited for the adventures that lay ahead on the open road.

Chapter 8: Design and Innovation

Emily Bennett's Influence on Car Design

As the 1960s transitioned into the 1970s, Emily Bennett's influence on car design became even more pronounced. Her commitment to blending form and function had already resulted in several successful models, and she was now poised to take Bennett Motors into a new era of design and innovation. Emily's unique vision and dedication to excellence ensured that Bennett cars would not only meet but exceed the expectations of an increasingly sophisticated consumer base.

Emily had always been fascinated by the interplay between aesthetics and engineering. She believed that a car should be a work of art as well as a feat of engineering, and this philosophy guided her design process. Her office was filled with sketches, models, and prototypes, each representing her pursuit of the perfect balance between beauty and functionality.

The Bennett Elegance: A Milestone in Design

In 1972, Bennett Motors introduced the Bennett Elegance, a mid-sized luxury sedan that would become one of Emily's most celebrated achievements. The Elegance was a testament to Emily's ability to merge cutting-edge technology with timeless design. It featured smooth, flowing lines, a spacious and meticulously crafted interior, and a range of advanced features that set it apart from its competitors.

The Elegance was designed with the modern driver in mind. It boasted a powerful yet efficient engine, responsive handling, and an array of safety features that underscored Bennett Motors' commitment to innovation. The interior was a marvel of ergonomic design, with every control and feature thoughtfully placed for maximum comfort and convenience.

The Bennett Elegance was an immediate success, receiving rave reviews from critics and consumers alike. It was praised for its elegant design, superior performance, and luxurious comfort. The car quickly became a favorite among executives, celebrities, and discerning drivers who valued both style and substance.

Innovative Features and Technology

Emily's vision for the Elegance extended beyond its exterior design. She was a strong advocate for incorporating the latest technology into Bennett cars, and the Elegance was no exception. The car featured several innovative features that were ahead of their time, setting new standards for the industry.

One of the standout features of the Elegance was its advanced climate control system. Emily understood the importance of comfort, and she worked closely with engineers to develop a system that allowed for precise temperature control, ensuring that passengers could enjoy a perfect environment regardless of the weather outside.

Another innovation was the Elegance's state-of-the-art sound system. Emily was a music enthusiast, and she wanted the car to provide an exceptional audio experience. The sound system, developed in collaboration with leading audio engineers, offered crystal-clear sound quality, allowing drivers and passengers to enjoy their favorite music like never before.

Safety was also a top priority for Emily. The Elegance included a reinforced frame, advanced braking system, and a suite of safety features designed to protect occupants in the event of an accident. These innovations not only enhanced the car's safety profile but also reinforced Bennett Motors' reputation for building reliable and secure vehicles.

Women in the Automotive Industry

Emily's success with the Bennett Elegance and other models helped pave the way for greater inclusion of women in the automotive industry. She became a role model and mentor to many aspiring female designers and engineers, demonstrating that talent and creativity were not confined by gender.

Emily frequently spoke at industry conferences and participated in design exhibitions, sharing her experiences and advocating for diversity and inclusion in the field. Her efforts helped change perceptions and opened doors for women in an industry that had traditionally been male-dominated.

Jack Bennett's Continued Impact

While Emily was revolutionizing car design, her brother Jack continued to make his mark in the world of motorsports. Jack's racing career, coupled with his technical expertise, played a crucial role in the development of high-performance vehicles for Bennett Motors. His insights and feedback were invaluable in refining the engineering and performance aspects of the company's cars.

Jack's involvement was particularly evident in the development of the Bennett Thunder, a high-performance sports car introduced in 1974. The Thunder was designed for speed enthusiasts and featured a powerful engine, aerodynamic design, and advanced suspension system. Jack tested the car extensively, pushing it to its limits on the track and providing crucial feedback to the engineering team.

The Bennett Thunder quickly gained a reputation for its exhilarating performance and superior handling. It became a favorite among racing enthusiasts and was featured in numerous car magazines and racing events. Jack's success on the track, combined with the Thunder's impressive capabilities, further cemented Bennett Motors' reputation for excellence.

Environmental Awareness and Sustainable Design

As the 1970s progressed, environmental concerns became increasingly prominent. The oil crisis of 1973 and growing awareness of air pollution highlighted the need for more sustainable automotive solutions. Bennett Motors, known for its innovation and adaptability, embraced this challenge and began exploring ways to reduce the environmental impact of its vehicles.

Thomas Bennett, Emily, and Jack were all committed to sustainability and believed that innovation was the key to addressing environmental challenges. They established a dedicated research and development team focused on creating more fuel-efficient engines, exploring alternative fuels, and developing new technologies to reduce emissions.

One of the notable outcomes of these efforts was the introduction of the Bennett Eco, a hybrid vehicle that combined a traditional gasoline engine with an electric motor. Launched in 1976, the Eco was one of the first hybrid cars on the market and represented a significant step forward in sustainable automotive design.

The Eco was praised for its fuel efficiency, reduced emissions, and innovative technology. It appealed to environmentally conscious consumers and demonstrated Bennett Motors' commitment to leading the industry toward a more sustainable future.

The Bennett Legacy: Innovation and Excellence

The 1970s were a decade of transformation and growth for Bennett Motors. The company's commitment to design excellence, technological innovation, and environmental sustainability set it apart from its competitors and ensured its continued success. The contributions of Thomas, Emily, and Jack Bennett were instrumental in shaping the company's direction and legacy.

Thomas Bennett's visionary leadership provided the foundation for the company's success. His dedication to quality and innovation inspired a culture of excellence that permeated every aspect of Bennett Motors. Emily's groundbreaking designs and commitment to integrating technology and aesthetics pushed the boundaries of what was possible in car design. Jack's passion for performance and racing expertise drove the development of high-performance vehicles that thrilled enthusiasts and set new standards for the industry.

As the decade drew to a close, Bennett Motors was well-positioned for the future. The lessons learned and the innovations developed during this period had laid the groundwork for continued success. The Bennett family's unwavering commitment to excellence, innovation, and sustainability ensured that Bennett Motors remained a leader in the automotive industry.

The story of Bennett Motors during the rise of muscle cars and the subsequent focus on design and sustainability highlights the company's ability to adapt and thrive in a rapidly changing world. The Bennett family's legacy of innovation and excellence continued to drive the company forward, setting the stage for future achievements and ensuring that Bennett Motors remained a symbol of quality and innovation in the automotive industry.

Chapter 9: Road Trips and Cultural Shifts

The Family Road Trip Tradition

The Bennett family's tradition of road trips had always been a source of inspiration and bonding. These journeys across America were more than vacations; they were integral to the Bennett way of life, providing opportunities to test new car models, gather consumer feedback, and explore the ever-changing landscape of the country. As the 1970s unfolded, these road trips became even more significant, reflecting the cultural shifts and transformations occurring in American society.

Thomas, Margaret, Emily, and Jack Bennett continued to take to the open road, each trip offering new adventures and insights. These journeys were meticulously planned, yet spontaneous enough to allow for unexpected detours and discoveries. The Bennetts believed that the best way to understand the needs and desires of their customers was to experience the road as they did, facing the same challenges and enjoying the same pleasures.

Journey Across the American Southwest

In the summer of 1972, the Bennetts embarked on a memorable road trip across the American Southwest. This journey was particularly significant, as it was the first major outing for the Bennett Elegance, Emily's masterpiece, and the company's latest luxury sedan. The goal was to test the Elegance in various conditions and to experience firsthand the unique landscapes and cultures of the Southwest.

The trip began in Detroit, with the family heading westward toward the vast deserts and rugged terrain of the Southwest. Their route took them through iconic locations such as the Grand Canyon, Monument Valley, and the deserts of New Mexico and Arizona. Each stop provided an opportunity to engage with locals, gather feedback on the

Elegance, and immerse themselves in the natural beauty and cultural richness of the region.

Exploring the Grand Canyon

The Grand Canyon was one of the highlights of the trip. As they approached the canyon's edge, the family was awestruck by the sheer scale and majesty of the landscape. They spent several days exploring the area, hiking along the rim, and venturing into the canyon itself. The Bennett Elegance performed admirably on the challenging roads leading to the Grand Canyon, showcasing its superior handling and comfort.

During their stay, Thomas and Emily took the opportunity to talk to other travelers, many of whom were curious about the sleek new sedan. The feedback they received was overwhelmingly positive, with many praising the Elegance's smooth ride, luxurious interior, and advanced features. This validation from everyday drivers reinforced their belief in the Elegance's design and functionality.

Monument Valley and Native American Culture

The journey continued to Monument Valley, a place known for its striking red rock formations and cultural significance. The Bennetts were deeply interested in learning about the history and traditions of the Navajo Nation, who called this area home. They spent time with Navajo guides, listening to stories and gaining a deeper appreciation for the land and its people.

Emily, always seeking inspiration for her designs, found herself particularly moved by the patterns and colors of Navajo art. She sketched extensively during their stay, capturing ideas that would later influence her future projects. The visit to Monument Valley was a profound experience, highlighting the importance of cultural respect and the richness of America's diverse heritage.

The Deserts of New Mexico and Arizona

As they traveled through the deserts of New Mexico and Arizona, the Bennetts encountered the challenges of extreme heat and rough terrain. The Bennett Elegance proved its resilience, handling the harsh conditions with ease. The car's advanced climate control system kept the interior cool and comfortable, even as temperatures soared outside.

The desert landscape, with its unique flora and fauna, provided a stark contrast to the bustling city life they had left behind. The family enjoyed the solitude and vastness of the desert, finding beauty in its simplicity and silence. These moments of quiet reflection were invaluable, allowing them to recharge and reconnect with each other.

Cultural Shifts and the Changing American Landscape

The 1970s were a time of significant cultural shifts in

America. The environmental movement was gaining momentum, with increasing awareness of the impact of human activities on the planet. The oil crisis of 1973 brought fuel efficiency to the forefront of consumer concerns, and there was a growing demand for cars that were both stylish and environmentally friendly.

Bennett Motors, always attuned to societal trends, responded to these changes with innovation and foresight. The Bennett Eco, introduced in 1976, was a direct response to the demand for more sustainable transportation options. This hybrid vehicle, combining a traditional gasoline engine with an electric motor, represented a significant technological advancement and a commitment to reducing the environmental footprint of automobiles.

Emily's Designs Reflecting Cultural Awareness

Emily Bennett's designs continued to evolve,

reflecting the cultural and environmental awareness of the era. She believed that cars could be both beautiful and responsible, and she worked tirelessly to integrate sustainable materials and technologies into her designs. The Bennett Eco was a prime example of this philosophy, featuring recycled materials, energy-efficient components, and a design that minimized environmental impact.

Emily also drew inspiration from the cultural diversity they encountered on their road trips. Her designs began to incorporate elements of the various cultures they had experienced, from the geometric patterns of Navajo art to the vibrant colors of Mexican textiles. These influences added a unique and personal touch to Bennett cars, resonating with consumers who appreciated both innovation and tradition.

Jack's Continued Passion for Racing

Jack Bennett's passion for racing remained

undiminished, and he continued to compete in events across the country. His experiences on the track provided valuable insights into high-performance engineering, which he shared with the Bennett Motors design team. Jack's input was instrumental in developing the Bennett Thunder, a sports car that combined speed, agility, and advanced technology.

The Thunder, introduced in 1974, was a testament to Jack's racing expertise and the company's commitment to performance. It featured a powerful engine, aerodynamic design, and cutting-edge suspension system, making it a favorite among racing enthusiasts and sports car aficionados. Jack's success on the track, combined with the Thunder's impressive capabilities, further solidified Bennett Motors' reputation for excellence.

Legacy and the Road Ahead

As the 1970s came to a close, Bennett Motors stood

at the forefront of the automotive industry, recognized for its innovation, design, and commitment to sustainability. The company's ability to adapt to changing cultural and environmental trends ensured its continued success and relevance in a rapidly evolving world.

The Bennett family's road trips, from the deserts of the Southwest to the bustling cities of America, had provided invaluable experiences and insights. These journeys reinforced the importance of understanding the needs and desires of their customers, staying connected to the cultural pulse of the nation, and continuously pushing the boundaries of what was possible in automotive design and technology.

The Bennett family's legacy of passion, creativity, and commitment to quality remained the cornerstone of the company's success. As they looked to the future, the Bennetts were ready to embrace new challenges and opportunities, guided by their unwavering commitment to excellence, sustainability, and innovation. The road ahead promised new adventures and discoveries, and the Bennett family was prepared to navigate it with the same spirit of curiosity and determination that had driven them for generations.

Chapter 10: The Digital Revolution

The Impact of Technology on Car Manufacturing

The 1980s and 1990s were transformative decades for the automotive industry, driven by rapid advancements in technology and a shift towards digitalization. Bennett Motors, always at the forefront of innovation, embraced these changes, integrating new technologies into their manufacturing processes and vehicle designs.

Thomas Bennett recognized early on that embracing digital technology was essential for maintaining a competitive edge. The company invested heavily in computer-aided design (CAD) and manufacturing (CAM) systems, revolutionizing the way cars were designed and produced. These technologies allowed for greater precision, efficiency, and flexibility in the manufacturing process, enabling Bennett Motors to produce higher-quality vehicles at a faster pace.

The use of robotics on the assembly line further enhanced production capabilities. Robots could perform repetitive tasks with unmatched accuracy and speed, reducing the potential for human error and increasing overall efficiency. This shift not only improved the quality of Bennett cars but also allowed the company to keep up with growing demand while maintaining stringent quality standards.

Megan Bennett's Investigative Journalism

By the late 1980s, a new generation of Bennetts was coming of age. Megan Bennett, the daughter of Jack Bennett, had inherited her family's passion for cars but chose to pursue a different path. With a keen interest in storytelling and a desire to explore the world, Megan became an investigative journalist, focusing on the automotive industry.

Megan's work often took her behind the scenes of car manufacturing plants, design studios, and racing tracks. She uncovered stories of innovation, explored the impact of technology on the industry, and highlighted the human stories behind the machines. Her articles and documentaries garnered critical acclaim and brought a fresh perspective to the world of automobiles.

In 1991, Megan's investigative work led her to a groundbreaking story about the rise of electric vehicles. She spent months researching and interviewing industry experts, engineers, and environmentalists. Her documentary, "Electric Dreams: The Future of Transportation," highlighted the potential of electric cars to revolutionize the industry and reduce environmental impact. The documentary's success brought widespread attention to the benefits of electric vehicles and positioned Megan as a leading voice in automotive journalism.

The Rise of Japanese Automakers

The late 20th century also saw the rise of Japanese automakers, who brought a new level of competition to the global automotive market. Companies like Toyota, Honda, and Nissan gained significant market share by offering reliable, fuel-efficient, and affordable vehicles. This shift forced established automakers, including Bennett Motors, to reevaluate their strategies and adapt to the changing landscape.

Thomas and Emily Bennett closely monitored these developments, recognizing the need to learn from their competitors. They studied Japanese manufacturing techniques, particularly the principles of lean manufacturing and just-in-time production. These methods emphasized efficiency, waste reduction, and continuous improvement, aligning well with Bennett Motors' commitment to innovation and quality.

The Bennett Envision: A New Era of Electric Vehicles

In response to the growing interest in sustainable transportation, Bennett Motors accelerated its research and development efforts in electric vehicle technology. The company's experience with the Bennett Eco had provided valuable insights, but Thomas and Emily knew that the next generation of electric cars needed to offer more in terms of range, performance, and affordability.

In 1996, Bennett Motors unveiled the Bennett Envision, an electric vehicle that represented a significant leap forward in automotive technology. The Envision featured a state-of-the-art battery system, providing an impressive range on a single charge. Its design combined aerodynamic efficiency with modern aesthetics, making it both practical and visually appealing.

The Envision's launch was met with enthusiasm from

both consumers and industry experts. It was praised for its innovative technology, environmental benefits, and driving experience. The car's success solidified Bennett Motors' position as a leader in the electric vehicle market and demonstrated the company's ability to adapt to evolving consumer preferences and technological advancements.

The Digital Dashboard: Integrating Technology into the Driving Experience

As digital technology became increasingly integrated into everyday life, consumers began to expect more from their vehicles. Bennett Motors responded by developing the Digital Dashboard, an advanced infotainment system that brought the latest technology to the driving experience. Launched in 1998, the Digital Dashboard featured a touch-screen interface, GPS navigation, real-time traffic updates, and seamless connectivity with smartphones and other devices.

The Digital Dashboard transformed the way drivers interacted with their cars, offering unprecedented levels of convenience, safety, and entertainment. It allowed drivers to access navigation, communication, and entertainment features with ease, enhancing the overall driving experience. The system also included advanced safety features, such as collision warnings and lane-keeping assistance, reflecting Bennett Motors' commitment to driver safety.

Emily Bennett played a key role in the development of the Digital Dashboard, working closely with software engineers and designers to ensure that the system was intuitive and user-friendly. Her emphasis on usability and aesthetics helped create a product that was both functional and appealing, setting a new standard for in-car technology.

Sustainability and Corporate Responsibility

Throughout the 1990s, Bennett Motors continued to

prioritize sustainability and corporate responsibility. The company implemented numerous initiatives to reduce its environmental impact, from improving manufacturing processes to promoting recycling and waste reduction. Bennett Motors also invested in renewable energy sources, such as solar and wind power, to supply its production facilities.

These efforts were part of a broader commitment to creating a sustainable future for the automotive industry. Bennett Motors believed that innovation and responsibility could go hand in hand, and the company's actions reflected this philosophy. By leading the way in sustainable practices, Bennett Motors not only enhanced its reputation but also contributed to the global effort to address environmental challenges.

Megan's Stories of Sustainability

Megan Bennett's journalism continued to play a

crucial role in highlighting the importance of sustainability in the automotive industry. She produced a series of articles and documentaries focusing on the environmental initiatives of major car manufacturers, including Bennett Motors. Her work shed light on the positive steps being taken and the challenges that remained.

One of her most impactful documentaries, "Driving Change: The Road to a Sustainable Future," explored the efforts of various automakers to reduce their carbon footprint and promote sustainable practices. The documentary featured interviews with industry leaders, environmentalists, and engineers, providing a comprehensive overview of the progress being made and the obstacles yet to be overcome.

Megan's storytelling resonated with audiences and helped raise awareness about the critical role of sustainability in the automotive industry. Her work not only informed but also inspired, encouraging consumers and manufacturers alike to prioritize environmental responsibility.

Global Expansion and Partnerships

As part of its strategy to promote sustainability on a global scale, Bennett Motors pursued partnerships and collaborations with other industry leaders, research institutions, and governments. These partnerships aimed to accelerate the development and adoption of green technologies and to share best practices across the industry.

One notable collaboration was with a leading battery manufacturer to develop next-generation battery technology with improved energy density, faster charging times, and longer lifespan. This partnership helped Bennett Motors stay at the forefront of electric vehicle innovation and ensure that its products remained competitive.

Bennett Motors also expanded its global footprint,

establishing manufacturing facilities and research centers in key international markets. These expansions not only supported local economies but also allowed the company to better understand and address the unique needs of different regions, further driving the adoption of sustainable transportation solutions worldwide.

Consumer Response and Market Trends

The early 2010s saw a significant shift in consumer attitudes towards sustainability. As awareness of climate change and environmental issues grew, more consumers began to prioritize eco-friendly products, including automobiles. Bennett Motors was well-positioned to capitalize on this trend, thanks to its long-standing commitment to green technology and sustainable practices.

Sales of the Bennett Fusion and Bennett Volt continued to grow, driven by their innovative features and environmental benefits. The company's focus on sustainability resonated with a new generation of consumers who were more conscious of their environmental impact and eager to support brands that aligned with their values.

Bennett Motors also expanded its product lineup to include a range of hybrid and electric vehicles, catering to different market segments and price points. This diversification helped the company attract a broader customer base and maintain its competitive edge in an increasingly crowded market.

Legacy and the Road Ahead

The digital revolution and the rise of sustainable transportation marked a significant chapter in the history of Bennett Motors. The company's commitment to innovation, technology, and environmental responsibility ensured its continued success and relevance in a rapidly changing world. Thomas, Emily, and Megan Bennett each played pivotal roles in navigating this transformative period, leveraging their unique skills and perspectives to drive the company forward.

The Bennett family's legacy of excellence and innovation continued to guide Bennett Motors into the future. As they looked ahead, they remained dedicated to pushing the boundaries of what was possible in automotive design and technology, striving to create a better, more sustainable world through their work. The road ahead was filled with promise, and the Bennett family was ready to embrace the challenges and opportunities that lay before them, continuing their journey of leadership and impact in the automotive industry.

Chapter 11: The Green Movement

Environmental Concerns and the Auto Industry

The early 2000s marked a significant turning point for the automotive industry as environmental concerns took center stage. Climate change, air pollution, and the depletion of natural resources became urgent global issues, prompting both consumers and regulators to demand more sustainable solutions. Bennett Motors, with its long-standing commitment to innovation and responsibility, was well-positioned to lead the way.

Thomas Bennett had always believed that the future of the automotive industry lay in sustainable practices. Under his guidance, Bennett Motors had already made significant strides in reducing emissions and improving fuel efficiency. However, he knew that more ambitious measures were needed to address the environmental challenges of the 21st century.

The Development of Hybrid and Electric Vehicles

Building on decades of research and development, the company focused on creating a new generation of hybrid and electric vehicles that offered superior performance, longer range, and lower environmental impact. Bennett Motors had pioneered hybrid technology with the introduction of the Bennett Eco in the 1970s, and this legacy continued into the new millennium.

In 2003, Bennett Motors unveiled the Bennett Fusion, a state-of-the-art hybrid vehicle that combined a highly efficient gasoline engine with a powerful electric motor. The Fusion represented a significant leap forward in hybrid technology, offering impressive fuel economy and reduced emissions without compromising on performance. The car featured regenerative braking, advanced battery management systems, and a seamless transition between gasoline and electric power.

The Fusion was met with enthusiasm from both consumers and environmental advocates. It quickly became a popular choice for eco-conscious drivers, earning accolades for its innovation and environmental benefits. The success of the Fusion reinforced Bennett Motors' reputation as a leader in green automotive technology and set the stage for the company's next major advancement.

The Launch of the Bennett Volt

In 2008, Bennett Motors introduced the Bennett Volt, a fully electric vehicle that promised to revolutionize the market. The Volt was the culmination of years of research and development, incorporating cutting-edge battery technology, lightweight materials, and advanced electronics. With a range of over 300 miles on a single charge, the Volt addressed one of the primary concerns of electric vehicle adoption: range anxiety.

The Volt's design was sleek and modern, reflecting

Bennett Motors' commitment to aesthetics and functionality. The car featured a spacious, high-tech interior with a digital dashboard, integrated infotainment system, and a host of smart features designed to enhance the driving experience. The Volt's battery could be charged using a standard household outlet or a dedicated charging station, making it convenient for everyday use.

The launch of the Bennett Volt was a landmark moment for the company. It was celebrated not only for its technological achievements but also for its potential to significantly reduce greenhouse gas emissions and dependence on fossil fuels. The Volt quickly gained a loyal following and became a symbol of Bennett Motors' vision for a sustainable future.

Megan Bennett's Stories of Sustainability

Megan Bennett continued her work as an investigative journalist, focusing on the intersection of technology and sustainability in the automotive industry. Her insightful reporting highlighted the challenges and opportunities facing automakers as they transitioned to greener technologies. Megan's articles and documentaries played a crucial role in raising public awareness and understanding of these issues.

One of Megan's most impactful projects during this period was a documentary titled "Electric Revolution: The Future of Driving." The film explored the development of electric vehicles, featuring interviews with engineers, environmentalists, and industry leaders, including her own family at Bennett Motors. The documentary provided an in-depth look at the technological advancements and the environmental benefits of electric vehicles, while also addressing the hurdles that still needed to be overcome.

Megan's storytelling was compelling and accessible, helping to demystify complex technologies and making the case for why electric vehicles were essential for a sustainable future. Her work earned critical acclaim and numerous awards, cementing her status as a leading voice in environmental journalism.

The Bennett Green Initiative

In 2010, Bennett Motors launched the Bennett Green Initiative, a comprehensive sustainability program aimed at reducing the company's environmental impact across all aspects of its operations. The initiative encompassed everything from manufacturing and supply chain management to product design and corporate practices.

Key components of the Bennett Green Initiative included:

Renewable Energy: Bennett Motors committed to powering its manufacturing plants and offices with 100% renewable energy by 2025. The company invested in solar panels, wind turbines, and other renewable energy sources to achieve this goal.

Sustainable Materials: The company increased its use of recycled and sustainable materials in vehicle production. This included everything from recycled metals and plastics to eco-friendly fabrics for car interiors.

Waste Reduction: Bennett Motors implemented zero-waste manufacturing processes, aiming to eliminate waste sent to landfills. This involved recycling, composting, and finding innovative ways to repurpose manufacturing by-products.

Corporate Responsibility: The initiative also focused on corporate social responsibility, including community engagement, employee wellness programs, and support for environmental conservation projects.

The Bennett Green Initiative was a bold and

ambitious plan that demonstrated the company's commitment to leading the industry in sustainability. It received widespread praise from environmental groups, consumers, and industry analysts, further enhancing Bennett Motors' reputation as a forward-thinking and responsible automaker.

Consumer Response and Market Trends

The early 2010s saw a significant shift in consumer attitudes towards sustainability. As awareness of climate change and environmental issues grew, more consumers began to prioritize eco-friendly products, including automobiles. Bennett Motors was well-positioned to capitalize on this trend, thanks to its long-standing commitment to green technology and sustainable practices.

Sales of the Bennett Fusion and Bennett Volt continued to grow, driven by their innovative features and environmental benefits. The company's focus on sustainability resonated with a new generation of consumers who were more conscious of their environmental impact and eager to support brands that aligned with their values.

Bennett Motors also expanded its product lineup to include a range of hybrid and electric vehicles, catering to different market segments and price points. This diversification helped the company attract a broader customer base and maintain its competitive edge in an increasingly crowded market.

Global Expansion and Partnerships

As part of its strategy to promote sustainability on a global scale, Bennett Motors pursued partnerships and collaborations with other industry leaders, research institutions, and governments. These partnerships aimed to accelerate the development and adoption of green technologies and to share best practices across the industry.

One notable collaboration was with a leading battery manufacturer to develop next-generation battery technology with improved energy density, faster charging times, and longer lifespan. This partnership helped Bennett Motors stay at the forefront of electric vehicle innovation and ensure that its products remained competitive.

Bennett Motors also expanded its global footprint, establishing manufacturing facilities and research centers in key international markets. These expansions not only supported local economies but also allowed the company to better understand and address the unique needs of different regions, further driving the adoption of sustainable transportation solutions worldwide.

The Bennett Family's Role

Thomas Bennett, now serving as Chairman Emeritus, remained actively involved in guiding the company's strategic direction. His vision and leadership continued to inspire the Bennett Motors team, and his commitment to sustainability set the tone for the company's future.

Emily Bennett, as CEO, was instrumental in driving the company's green initiatives. Her innovative designs and emphasis on sustainable materials and practices played a crucial role in the success of the Bennett Fusion and Bennett Volt. Emily's leadership ensured that Bennett Motors remained at the forefront of automotive innovation and sustainability.

Jack Bennett, with his extensive experience in high-performance engineering, focused on integrating green technologies into the company's performance models. He believed that sustainability and performance were not mutually exclusive and worked to develop electric and hybrid sports cars that delivered both excitement and efficiency.

Megan Bennett's investigative journalism continued to shine a spotlight on the automotive industry's evolution. Her articles and documentaries provided valuable perspectives on the impact of new technologies, the importance of sustainability, and the human stories behind the machines. Megan's work helped bridge the gap between the industry and the public, fostering a greater understanding and appreciation of the challenges and opportunities ahead.

Looking to the Future

As Bennett Motors looked to the future, the company remained committed to driving innovation, sustainability, and excellence in the automotive industry. The green movement had set the stage for new possibilities, and Bennett Motors was ready to continue its journey of leadership and innovation, paving the way for a brighter and more sustainable future.

The Bennett family's dedication to innovation, sustainability, and excellence remained the cornerstone of the company's success. Their collective efforts ensured that Bennett Motors remained a leader in the rapidly evolving automotive industry, setting new standards for performance, sustainability, and corporate responsibility.

With a clear vision and a commitment to their core values, the Bennetts were ready to embrace the future, confident that their legacy of excellence and innovation would continue to drive Bennett Motors forward. The road ahead was filled with promise, and the Bennett family was prepared to navigate it with the same passion, creativity, and determination that had defined their journey for generations.

Chapter 12: Autonomous and Connected Vehicles

The Future of Driving

As the 2010s progressed, the automotive industry began to witness a revolution driven by the development of autonomous and connected vehicle technologies. The prospect of self-driving cars and vehicles capable of communicating with each other and with infrastructure promised to transform transportation, making it safer, more efficient, and more accessible. Bennett Motors, always at the forefront of innovation, embraced these new technologies, seeing them as the next frontier in automotive excellence.

Thomas Bennett, nearing the end of his illustrious career, recognized the profound impact these technologies would have on the industry. He encouraged Emily, Jack, and the rest of the Bennett

Motors team to explore the potential of autonomous and connected vehicles, investing in research and development and forming strategic partnerships to advance their efforts.

Technological Challenges and Opportunities

Developing autonomous vehicles presented numerous challenges, from perfecting the technology to ensuring safety and regulatory compliance. However, it also offered unprecedented opportunities to redefine the driving experience and address many of the issues plaguing traditional transportation systems, such as traffic congestion, accidents, and environmental impact.

Emily Bennett spearheaded the company's efforts in autonomous vehicle development. She collaborated with leading technology companies, universities, and research institutions to integrate the latest advancements in artificial intelligence (AI), machine learning, and sensor technology into Bennett's vehicle designs. Her goal was to create a fleet of self-driving cars that were not only safe and reliable but also luxurious and enjoyable to drive.

Jack Bennett, with his extensive experience in high-performance engineering, focused on the integration of autonomous technology into sports cars. He believed that autonomous driving should not come at the expense of excitement and performance. Jack's team worked on developing systems that could enhance the driving experience, allowing drivers to switch between autonomous and manual modes seamlessly.

The Bennett Liberty: A Leap into Autonomy

In 2020, Bennett Motors unveiled the Bennett Liberty, its first fully autonomous vehicle. The Liberty represented a significant milestone in the company's history, showcasing years of research, development, and collaboration. The vehicle was designed to operate safely and efficiently without human intervention, thanks to a sophisticated array of sensors, cameras, radar, and lidar systems that provided a 360-degree view of the surroundings.

The Liberty's AI-powered system could process vast amounts of data in real-time, making split-second decisions to navigate traffic, avoid obstacles, and ensure the safety of its passengers. The car was also equipped with advanced connectivity features, allowing it to communicate with other vehicles, traffic signals, and smart infrastructure to optimize routes and reduce travel time.

Inside, the Bennett Liberty offered a luxurious and comfortable environment, reflecting Emily's design philosophy. The spacious interior featured high-quality materials, customizable lighting, and a state-of-the-art infotainment system. Passengers could relax, work, or enjoy entertainment while the car handled the driving. The Liberty's launch was a resounding success, attracting widespread attention and acclaim from both the automotive industry and the tech world.

Connected Vehicles and Smart Infrastructure

The development of connected vehicle technology went hand-in-hand with the rise of autonomous driving. Connected vehicles could share information with each other and with smart infrastructure, creating a more efficient and safer transportation network. Bennett Motors recognized the potential of this technology to transform urban mobility and reduce the environmental impact of transportation.

Emily and her team worked on integrating advanced connectivity features into Bennett's vehicle lineup. These features included vehicle-to-vehicle (V2V) and vehicle-to-infrastructure (V2I) communication systems, which allowed cars to share data about traffic conditions, road hazards, and optimal routes. This connectivity helped reduce traffic congestion, improve fuel efficiency, and enhance overall safety.

Bennett Motors also collaborated with cities and governments to develop smart infrastructure projects. These projects involved deploying sensors and communication networks on roads, traffic lights, and other infrastructure to create a seamless and intelligent transportation system. The goal was to enable vehicles and infrastructure to work together harmoniously, optimizing traffic flow and reducing the risk of accidents.

Safety and Regulatory Compliance

Ensuring the safety and regulatory compliance of autonomous and connected vehicles was a top priority for Bennett Motors. The company worked closely with regulators, industry groups, and safety organizations to develop and adhere to stringent safety standards. Bennett's autonomous vehicles underwent rigorous testing in various conditions to validate their performance and reliability.

Emily Bennett played a key role in advocating for clear and consistent regulatory frameworks that would support the safe deployment of autonomous vehicles. She participated in industry panels, testified before government bodies, and collaborated with other automakers to promote best practices and share knowledge. Her efforts helped shape the regulatory landscape, paving the way for the broader adoption of autonomous and connected vehicle technologies.

Consumer Adoption and Trust

Gaining consumer trust and acceptance was another critical aspect of introducing autonomous vehicles. Many people were excited about the potential benefits of self-driving cars, but concerns about safety, reliability, and privacy remained. Bennett Motors launched an extensive public education campaign to address these concerns and build confidence in their autonomous technology.

The campaign included demonstrations, test drives, and informative content that explained how autonomous systems worked and the measures taken to ensure safety. Bennett Motors also offered training programs for drivers to familiarize them with the new technology and its features. These efforts helped demystify autonomous driving and encouraged consumers to embrace the future of transportation.

Megan Bennett's Investigative Journalism

Megan Bennett continued to play a crucial role in documenting and explaining the advancements in autonomous and connected vehicle technology. Her investigative work provided an in-depth look at the development, testing, and deployment of these technologies, highlighting both the potential benefits and the challenges.

One of her notable documentaries during this period, titled "The Road Ahead: Navigating the Autonomous Revolution," explored the impact of self-driving cars on society. The film featured interviews with engineers, policymakers, and everyday drivers, providing a comprehensive overview of the technology's promise and the obstacles that still needed to be overcome. Megan's balanced and insightful reporting helped inform public discourse and foster a better understanding of autonomous vehicles.

The Bennett Legacy: Pioneering the Future

The introduction of the Bennett Liberty and the company's advancements in connected vehicle technology marked a new chapter in Bennett Motors' storied history. The company's commitment to innovation, safety, and sustainability continued to drive its success, ensuring that Bennett Motors remained a leader in the rapidly evolving automotive industry.

Thomas Bennett, reflecting on his career, felt immense pride in the accomplishments of his children and the company he had built. Emily's visionary designs, Jack's engineering prowess, and Megan's impactful journalism had all contributed to Bennett Motors' legacy of excellence and innovation.

As the company looked to the future, it remained dedicated to pushing the boundaries of what was possible in automotive technology. The rise of autonomous and connected vehicles represented just one of many steps on Bennett Motors' journey to create a safer, more efficient, and sustainable transportation system.

The introduction of autonomous and connected vehicles was a testament to the Bennett family's vision and determination. Their efforts ensured that Bennett Motors remained at the forefront of technological advancements, setting new standards for safety, performance, and sustainability. The

Bennetts' legacy of innovation and leadership continued to drive the company forward, shaping the future of transportation and solidifying their place in automotive history.

Chapter 13: Reflecting on a Century of Change

The Evolution of Automotive Design

As the 2020s progressed, Bennett Motors reached a significant milestone: the centennial of its founding. Celebrating 100 years of innovation, the company took time to reflect on its journey, the evolution of automotive design, and the profound impact it had on the industry and society. The centennial was not just a celebration of the past but also a moment to envision the future.

Emily Bennett, now the CEO of Bennett Motors, led the company through this reflective period. She organized a series of events, exhibitions, and publications to commemorate the company's rich history. These initiatives highlighted the key milestones, iconic models, and technological breakthroughs that had defined Bennett Motors over the decades.

Lessons Learned from the Past

One of the central themes of the centennial celebration was the importance of learning from the past to inform the future. The company's history was filled with stories of resilience, innovation, and adaptation. From surviving the Great Depression and World War II to leading the charge in sustainable and autonomous vehicle technologies, Bennett Motors had consistently demonstrated its ability to navigate challenges and seize opportunities.

Thomas Bennett's early vision of quality, innovation, and customer-centric design laid the foundation for the company's enduring success. His belief in the power of technology to transform society and improve lives resonated through the generations, guiding the company's evolution. Emily emphasized that these core values remained as relevant today as they were a century ago.

The Importance of Heritage

The centennial celebrations included a retrospective exhibition at the Bennett Motors Museum, showcasing the company's most iconic models, from the early Bennett Roadster to the latest Bennett Liberty. The exhibition also featured personal artifacts, design sketches, and historical documents that told the story of the Bennett family and their passion for automobiles.

Visitors to the exhibition were treated to an immersive experience, with interactive displays and virtual reality tours that brought the company's history to life. The exhibit highlighted the technological advancements and design philosophies that had defined each era, illustrating how Bennett Motors had consistently pushed the boundaries of what was possible in automotive engineering.

A Vision for the Future

While reflecting on the past, the centennial celebration also looked forward to the future. Emily and her team presented their vision for the next century of Bennett Motors, emphasizing a continued commitment to innovation, sustainability, and excellence. They outlined strategic initiatives aimed at addressing the evolving needs of consumers and the challenges of a rapidly changing world.

One of the key components of this vision was the expansion of Bennett Motors' electric and autonomous vehicle lineup. Building on the success of the Bennett Volt and Bennett Liberty, the company planned to introduce a range of new models that combined cutting-edge technology with sustainable design. These vehicles would feature advanced battery technology, enhanced connectivity, and innovative safety features, setting new standards for performance and environmental responsibility.

Sustainability and Corporate Responsibility

Sustainability remained a central focus of Bennett Motors' future strategy. The company committed to achieving carbon neutrality by 2035, implementing comprehensive measures to reduce emissions, improve energy efficiency, and promote circular economy practices. This included increasing the use of recycled materials, optimizing manufacturing processes, and investing in renewable energy sources.

Bennett Motors also pledged to enhance its corporate social responsibility efforts, supporting community development, education, and environmental conservation projects. Emily believed that the company had a responsibility to contribute positively to society and to help address some of the most pressing challenges of our time.

Innovation and Collaboration

Innovation continued to be the driving force behind Bennett Motors' success. The company planned to expand its research and development efforts, exploring new technologies and materials that could revolutionize the automotive industry. This included advancements in battery technology, lightweight materials, and artificial intelligence.

Collaboration was another key element of Bennett Motors' future strategy. The company sought to forge new partnerships with technology firms, research institutions, and other automakers to drive innovation and share best practices. Emily believed that collaboration was essential for tackling the complex challenges of the future and for achieving the company's ambitious goals.

The Bennett Legacy Continues

As the centennial celebrations came to a close, the

Bennett family reflected on their journey and the legacy they had built. Thomas Bennett, now retired but still actively involved in an advisory capacity, felt immense pride in what his family and company had achieved. He saw the same passion and commitment in Emily, Jack, and Megan that had driven him throughout his career.

Jack Bennett, still passionate about racing and high-performance engineering, continued to play a key role in the company's development of sports cars and performance vehicles. His insights and expertise were invaluable in ensuring that Bennett Motors remained a leader in this segment.

Megan Bennett's journalism continued to shine a spotlight on the automotive industry's evolution. Her articles and documentaries provided valuable perspectives on the impact of new technologies, the importance of sustainability, and the human stories behind the machines. Megan's work helped bridge the gap between the industry and the public, fostering a greater understanding and appreciation of the challenges and opportunities ahead.

Global Impact and Reach

Bennett Motors' centennial celebrations also highlighted the company's global impact and reach. From its humble beginnings in Detroit, Bennett Motors had grown into a global brand, with a presence in markets around the world. The company's commitment to quality, innovation, and sustainability resonated with consumers and partners across different cultures and regions.

Emily and her team emphasized the importance of understanding and respecting the diverse needs and preferences of their global customers. They committed to designing and manufacturing vehicles that met the highest standards of quality and safety, while also reflecting the unique characteristics of each market.

Embracing New Technologies

As part of their forward-looking strategy, Bennett Motors invested heavily in emerging technologies that promised to reshape the automotive landscape. The company's research and development efforts focused on several key areas:

Advanced Battery Technology: Building on the success of their electric vehicles, Bennett Motors aimed to develop next-generation batteries with higher energy density, faster charging times, and longer lifespans. These advancements would further enhance the performance and convenience of electric vehicles, making them an even more attractive option for consumers.

Autonomous Driving Systems: Bennett Motors continued to refine their autonomous driving technology, aiming to make self-driving cars more reliable and accessible. This included improving AI algorithms, enhancing sensor accuracy, and ensuring that autonomous vehicles could safely navigate complex urban environments.

Artificial Intelligence and Machine Learning: AI and machine learning played a crucial role in developing smarter, more efficient vehicles. Bennett Motors invested in these technologies to enhance various aspects of vehicle design, from predictive maintenance systems to personalized driver experiences.

Sustainable Materials: The company explored new materials that were both lightweight and environmentally friendly. This included researching bio-based plastics, recycled composites, and other innovative materials that could reduce the environmental impact of vehicle production.

Corporate Social Responsibility and Community Engagement

Bennett Motors also strengthened its commitment to

corporate social responsibility (CSR) and community engagement. The company launched several initiatives aimed at supporting local communities, promoting education, and addressing environmental challenges. These efforts included:

Community Development: Bennett Motors partnered with local organizations to support community development projects, such as building affordable housing, improving infrastructure, and creating green spaces. These projects aimed to enhance the quality of life for residents and foster a sense of community.

Education and Training: The company invested in educational programs that encouraged young people to pursue careers in science, technology, engineering, and mathematics (STEM). This included providing scholarships, sponsoring STEM competitions, and partnering with schools to offer hands-on learning experiences.

Environmental Conservation: Bennett Motors supported various environmental conservation initiatives, such as reforestation projects, wildlife protection programs, and clean water initiatives. These efforts aimed to preserve natural habitats and promote biodiversity.

A New Generation of Leaders

As Bennett Motors looked to the future, the company recognized the importance of nurturing the next generation of leaders. Emily, Jack, and Megan Bennett played a crucial role in mentoring young talent and fostering a culture of innovation and excellence. They identified promising individuals within the company and provided them with the opportunities and resources needed to develop their skills and leadership potential.

One of these rising stars was Alex Bennett, Jack's son, who had recently joined the company after completing his engineering degree. Alex shared his father's passion for high-performance vehicles and brought fresh ideas and perspectives to the team. He quickly made a name for himself with his innovative approach to vehicle design and engineering.

Under the guidance of Emily, Jack, and Megan, Alex and other young leaders were poised to carry forward the legacy of Bennett Motors. Their energy, creativity, and commitment to excellence ensured that the company would continue to thrive and innovate in the years to come.

The Road Ahead

As Bennett Motors celebrated its centennial, the company stood at the forefront of the automotive industry, poised to continue its legacy of innovation, sustainability, and excellence. The Bennett family's vision and dedication had transformed a small workshop into a global brand, and their commitment to pushing the boundaries of what was possible in automotive design and technology remained as strong as ever.

The journey of Bennett Motors was a testament to the power of innovation, resilience, and family. As they looked to the future, the Bennetts were ready to embrace new challenges and opportunities, confident that their legacy of excellence would guide them on the road ahead. The next century promised to be as transformative and inspiring as the first, driven by the same passion and vision that had defined Bennett Motors from the beginning.

Chapter 14: Passing the Torch

Preparing the Next Generation

As Bennett Motors celebrated its centennial, the company also focused on preparing for the future by nurturing the next generation of leaders. Thomas Bennett, having guided the company through decades of growth and innovation, understood the importance of succession planning. He believed that the values and vision that had driven Bennett Motors for a century should be preserved and advanced by future generations.

Emily Bennett, now the CEO, took on the responsibility of mentoring young talent within the company. She established the Bennett Leadership Development Program, designed to identify and cultivate promising individuals who could lead Bennett Motors into the future. This program offered comprehensive training in various aspects of the automotive industry, from engineering and design to marketing and management.

One of the standout participants in the program was Alex Bennett, Jack's son. With a degree in mechanical engineering and a passion for high-performance vehicles, Alex quickly distinguished himself as a talented and innovative engineer. Emily saw in him the same drive and dedication that had propelled her father and herself, and she took a personal interest in his development.

Innovations for the Future

Under Emily's leadership, Bennett Motors continued to push the boundaries of innovation. The company expanded its research and development efforts, focusing on emerging technologies that promised to revolutionize the automotive industry. This included advancements in electric vehicles, autonomous driving, and connected car technologies.

One of the key projects was the development of the

Bennett Aurora, an electric vehicle that combined cutting-edge technology with unparalleled luxury and performance. The Aurora featured an advanced battery system that offered an extended range, rapid charging capabilities, and a suite of smart features that enhanced the driving experience. The car's sleek design and state-of-the-art interior reflected Emily's commitment to blending form and function.

Emily also prioritized sustainability, ensuring that the Aurora was manufactured using environmentally friendly materials and processes. This focus on sustainability resonated with consumers and reinforced Bennett Motors' reputation as a leader in green automotive technology.

Jack Bennett's Role in Performance Vehicles

Jack Bennett, known for his passion for racing and high-performance engineering, continued to play a crucial role in the development of performance vehicles at Bennett Motors. He spearheaded the creation of the Bennett Vortex, a high-performance sports car that embodied the spirit of speed and innovation. The Vortex featured a powerful electric drivetrain, aerodynamic design, and advanced handling systems that made it a formidable competitor on the track.

Jack's involvement ensured that Bennett Motors maintained its reputation for producing high-performance vehicles that thrilled enthusiasts and set new standards in the industry. His hands-on approach and attention to detail were instrumental in refining the Vortex's design and performance, making it a standout model in Bennett Motors' lineup.

Megan Bennett's Continued Impact

Megan Bennett's investigative journalism continued to provide valuable insights into the automotive industry. Her work highlighted the technological advancements, environmental initiatives, and human stories that defined the industry's evolution. Megan's documentaries and articles were widely regarded as authoritative sources of information and analysis, helping to shape public understanding and policy discussions.

Megan's latest project, a documentary titled "Driving Tomorrow: The Future of Automobiles," explored the transformative impact of electric and autonomous vehicles on society. The film featured interviews with industry leaders, policymakers, and everyday drivers, offering a comprehensive look at the opportunities and challenges of the automotive revolution. Megan's storytelling captured the excitement and complexity of this pivotal moment in automotive history.

The Importance of Mentorship and Legacy

Emily recognized the importance of mentorship in preserving the legacy of Bennett Motors. She established a mentorship program within the company, pairing experienced leaders with young professionals. This initiative aimed to foster a culture of learning and collaboration, ensuring that the company's values and knowledge were passed down to the next generation.

The mentorship program also focused on promoting diversity and inclusion within the company. Emily believed that a diverse workforce was essential for driving innovation and addressing the complex challenges of the future. The program provided opportunities for employees from different backgrounds and perspectives to contribute to the company's success and develop their careers.

Innovative Technologies and Sustainability

As Bennett Motors looked to the future, the company remained committed to sustainability and environmental responsibility. The development of the Bennett Aurora and other electric vehicles was part of a broader strategy to reduce the company's carbon footprint and promote sustainable transportation solutions. Bennett Motors also invested in renewable energy sources, eco-friendly manufacturing processes, and recycling initiatives.

One of the company's most ambitious projects was the creation of the Bennett Sustainability Campus, a state-of-the-art research and development facility dedicated to advancing sustainable automotive technologies. The campus included labs for battery research, autonomous driving systems, and connected vehicle technologies, as well as facilities for testing and prototyping new models.

The Bennett Sustainability Campus was a testament to the company's commitment to leading the industry in sustainability and innovation. It provided a collaborative environment where engineers, designers, and researchers could work together to develop the next generation of sustainable vehicles.

Corporate Social Responsibility and Community Engagement

Bennett Motors also expanded its corporate social responsibility efforts, focusing on community engagement and support for social and environmental causes. The company launched the Bennett Foundation, which funded initiatives in education, environmental conservation, and community development. The foundation supported projects that aligned with the company's values and made a positive impact on society.

Emily, Jack, and Megan were actively involved in the foundation's work, participating in community events and initiatives. They believed that giving back to society was an essential part of the company's mission and legacy. The foundation's projects included scholarships for students pursuing careers in engineering and technology, grants for environmental conservation efforts, and partnerships with local organizations to support community development.

Passing the Torch: The Next Generation of Leaders

As the centennial celebrations drew to a close, Emily, Jack, and Megan reflected on their family's legacy and the future of Bennett Motors. They were committed to ensuring that the company remained a leader in innovation, sustainability, and excellence. To achieve this, they focused on preparing the next generation of leaders who would carry forward the values and vision that had defined Bennett Motors for a century.

Emily identified several young executives who showed exceptional promise and potential. She worked closely with them, providing mentorship and guidance to help them develop their skills and leadership capabilities. These rising stars were given opportunities to lead key projects, participate in strategic planning, and gain experience in different areas of the company.

One of these young leaders was Alex Bennett, the son of Jack Bennett. Alex had grown up immersed in the world of cars and racing, inheriting his father's passion for high-performance engineering. After earning a degree in mechanical engineering, Alex joined Bennett Motors and quickly made a name for himself with his innovative ideas and dedication to excellence.

Alex was instrumental in the development of the Bennett Falcon, a next-generation electric sports car that combined cutting-edge technology with stunning design. The Falcon featured an advanced electric drivetrain, lightweight materials, and a range of smart features that enhanced performance and driving experience. Alex's work on the Falcon earned him recognition as one of the industry's rising stars.

Looking to the Future

As Bennett Motors celebrated its centennial, the company stood at the forefront of the automotive industry, poised to continue its legacy of innovation, sustainability, and excellence. The Bennett family's vision and dedication had transformed a small workshop into a global brand, and their commitment to pushing the boundaries of what was possible in automotive design and technology remained as strong as ever.

The journey of Bennett Motors was a testament to the power of innovation, resilience, and family. As they looked to the future, the Bennetts were ready to embrace new challenges and opportunities, confident that their legacy of excellence would guide them on the road ahead. The next century promised to be as transformative and inspiring as the first, driven by the same passion and vision that had defined Bennett Motors from the beginning.

Chapter 15: Embracing the Future

Technological Convergence and the New Era

As the 2030s approached, Bennett Motors found itself at the intersection of several transformative technologies. The convergence of electric propulsion, autonomous driving, and connected vehicle systems was revolutionizing the automotive industry. For Bennett Motors, this new era represented both a challenge and an opportunity to redefine transportation once again.

Emily Bennett, having guided the company through decades of innovation, was determined to lead Bennett Motors into this new era. She believed that the future of transportation lay in creating a seamless, integrated experience that combined the latest technological advancements with the company's longstanding commitment to quality and innovation.

The Bennett Nexus: A Vision of the Future

In 2032, Bennett Motors unveiled the Bennett Nexus, a groundbreaking vehicle that embodied the company's vision for the future. The Nexus was a fully electric, autonomous, and connected car designed to provide an unparalleled driving experience. It featured advanced AI systems, state-of-the-art battery technology, and a level of connectivity that transformed the vehicle into a mobile hub.

The Nexus's AI system, named "EVE" (Enhanced Vehicle Experience), could learn and adapt to the driver's preferences, providing personalized recommendations and optimizing driving conditions. EVE could manage everything from route planning to in-car entertainment, ensuring that every journey was both efficient and enjoyable.

The vehicle's connectivity features allowed it to communicate seamlessly with other cars, infrastructure, and smart devices. This enabled real-time traffic management, predictive maintenance, and a host of other services that enhanced convenience and safety. The Nexus was also equipped with advanced safety features, including a comprehensive sensor suite and redundant systems to ensure reliability and security.

Sustainability and Eco-Friendly Initiatives

The Bennett Nexus was designed with sustainability at its core. The vehicle utilized lightweight, recycled materials, and its manufacturing process minimized waste and energy consumption. The Nexus's battery system, developed in partnership with leading energy companies, offered unprecedented range and efficiency, with rapid charging capabilities that made long-distance travel practical and convenient.

In addition to the Nexus, Bennett Motors launched a series of eco-friendly initiatives aimed at reducing the company's overall environmental impact. This included expanding their renewable energy usage, implementing circular economy principles, and investing in carbon offset projects. These efforts were part of a broader commitment to achieving carbon neutrality by 2035.

The Role of Artificial Intelligence

Artificial intelligence played a crucial role in the development and operation of Bennett's new generation of vehicles. Emily's daughter, Claire Bennett, who had recently joined the company as the head of AI development, was instrumental in advancing these technologies. Claire's expertise in AI and machine learning brought a new dimension to Bennett Motors' innovation efforts.

Claire's team focused on developing AI systems that could enhance both the driving experience and the vehicle's performance. This included creating algorithms that optimized energy usage, improved autonomous driving capabilities, and provided real-time data analytics. Claire also worked on integrating AI with Bennett Motors' manufacturing processes, using machine learning to increase efficiency and quality control.

Global Expansion and Market Adaptation

As Bennett Motors continued to innovate, the company also expanded its global footprint. New manufacturing plants and research centers were established in strategic locations worldwide, allowing Bennett Motors to better serve diverse markets and adapt to local preferences. This global expansion was accompanied by partnerships with regional technology firms and universities, fostering a collaborative approach to innovation.

Emily and her team emphasized the importance of understanding and respecting cultural differences in their global operations. They conducted extensive market research and engaged with local communities to ensure that Bennett Motors' vehicles and services met the unique needs and expectations of customers around the world.

The Bennett Family's Ongoing Legacy

Thomas Bennett, now enjoying his retirement, watched with pride as his family continued to drive the company forward. His grandchildren, including Alex and Claire, brought fresh perspectives and new ideas to Bennett Motors, ensuring that the company remained at the cutting edge of the industry.

Alex Bennett, following in his father Jack's footsteps, focused on the performance aspects of Bennett Motors' new electric and autonomous vehicles. His work on the Bennett Falcon, a high-performance electric sports car, demonstrated that sustainability and excitement could go hand-in-hand. The Falcon featured a powerful electric drivetrain, advanced aerodynamics, and a sleek design that made it a favorite among car enthusiasts.

Megan Bennett, through her continued work in investigative journalism, kept a close eye on industry developments and the societal impacts of new technologies. Her documentaries and articles provided valuable insights into the ethical, environmental, and social implications of the automotive revolution, helping to shape public discourse and policy.

Challenges and Opportunities

Despite the excitement and promise of new technologies, Bennett Motors faced significant challenges. The rapidly changing regulatory landscape, cybersecurity threats, and the need to maintain consumer trust were constant concerns. Emily, Claire, and the rest of the leadership team were committed to addressing these challenges head-on, leveraging their expertise and the company's resources to navigate the complex environment.

One of the key challenges was ensuring the security and privacy of connected and autonomous vehicles. Claire's team worked tirelessly to develop robust cybersecurity measures, protecting the vehicles and their users from potential threats. This included implementing advanced encryption, intrusion detection systems, and regular security audits.

Another challenge was the ethical considerations surrounding AI and autonomous driving. Bennett Motors established an ethics committee, comprising experts from various fields, to guide the development and deployment of their technologies. This committee worked to ensure that Bennett Motors' innovations were aligned with ethical standards and contributed positively to society.

A Vision for the Next Century

As Bennett Motors looked to the future, the company remained committed to its founding principles of innovation, quality, and customer focus. Emily, Alex, Claire, and Megan were united in their vision of creating a sustainable, connected, and autonomous future for transportation. They believed that Bennett Motors could continue to lead the industry by embracing change, fostering collaboration, and prioritizing the needs and values of their customers.

The Bennett family's legacy of excellence and innovation continued to inspire the entire organization. With a clear vision and a dedicated team, Bennett Motors was well-positioned to navigate the challenges and opportunities of the 21st century, ensuring that the company remained a leader in the automotive industry for generations to come.

The road ahead was filled with promise, and the Bennett family was ready to drive into the future, confident in their ability to shape the next chapter of automotive history.

Epilogue: A Legacy of Innovation

Reflecting on a Century

As Bennett Motors celebrated its centennial milestone, the company and the Bennett family took a moment to reflect on their extraordinary journey. From its humble beginnings in a small workshop in Detroit to becoming a global leader in automotive innovation, Bennett Motors had traversed a path marked by resilience, creativity, and a relentless pursuit of excellence.

Thomas Bennett, the visionary founder, had set the tone with his unwavering commitment to quality and innovation. His legacy was carried forward by his children and grandchildren, each contributing their unique talents and perspectives to shape the company's evolution. The centennial celebration was not just a look back at the past but a reaffirmation of the values that had driven Bennett Motors for a century.

Family Reflections

Thomas, now enjoying his retirement, spent much of his time reflecting on the company's history and the accomplishments of his family. Watching his children and grandchildren continue the work he had started filled him with immense pride. He saw in them the same passion and determination that had driven him throughout his career.

Emily Bennett, as CEO, had steered the company through some of its most transformative years. Her innovative designs and leadership had positioned Bennett Motors as a pioneer in electric and autonomous vehicle technology. Under her guidance, the company had not only embraced change but had also led the industry in sustainability and technological advancements.

Jack Bennett, with his deep love for performance vehicles, had ensured that Bennett Motors remained at the forefront of high-performance engineering. His contributions to the development of electric sports cars like the Bennett Falcon demonstrated that sustainability and excitement could coexist, delighting car enthusiasts around the world.

Megan Bennett, through her impactful journalism, had provided a critical lens on the industry, highlighting both its triumphs and challenges. Her work had fostered greater public understanding and dialogue about the future of transportation, ensuring that ethical considerations and social impacts were part of the conversation.

Claire Bennett, the new head of AI development, represented the future of Bennett Motors. Her expertise in artificial intelligence and her innovative approach to integrating AI with automotive technology promised to keep the company at the cutting edge of the industry. Her leadership was instrumental in the development of the Bennett Nexus, a vehicle that set new standards for autonomy, connectivity, and sustainability.

A Commitment to the Future

As the Bennett family looked ahead, they remained steadfast in their commitment to driving innovation and sustainability in the automotive industry. The challenges of the 21st century—climate change, technological disruption, and evolving consumer expectations—required bold thinking and decisive action. Bennett Motors was prepared to meet these challenges head-on, guided by the same principles that had driven its success for the past century.

The Bennett Sustainability Campus was a testament to this commitment. This state-of-the-art research and development facility was dedicated to advancing green technologies and promoting sustainable practices across the automotive industry. It symbolized the company's dedication to creating a better, more sustainable future.

Community and Global Impact

Bennett Motors' impact extended beyond its technological innovations. The company's corporate social responsibility initiatives and community engagement efforts had made a significant difference in the lives of many. Through the Bennett Foundation, the company supported educational programs, environmental conservation projects, and community development initiatives around the world.

Emily, Jack, Megan, and Claire were all actively involved in these efforts, believing that giving back to society was an integral part of the company's mission. Their work in these areas helped build stronger communities and promoted a more inclusive and sustainable world.

Looking Forward

The centennial celebration was both a reflection on

an illustrious past and a springboard into an exciting future. Bennett Motors, under the leadership of the Bennett family, was poised to continue its legacy of innovation, quality, and excellence. The road ahead was filled with possibilities, and the company was ready to navigate the challenges and opportunities that lay before it.

As the automotive industry continued to evolve, Bennett Motors remained committed to its founding principles. The company's history was a testament to the power of innovation, resilience, and family. With a clear vision and a dedicated team, Bennett Motors was well-equipped to lead the industry into the next century.

The Bennett family's legacy was one of constant evolution and unwavering dedication to excellence. As they embraced the future, they carried with them the lessons of the past and the passion that had always driven their success. The journey of Bennett Motors was far from over, and the next chapter promised to be as transformative and inspiring as the first.

The Bennett family was ready to continue their journey, confident in their ability to shape the future of transportation and ensure that Bennett Motors remained a symbol of innovation, sustainability, and excellence for generations to come.

About the Author

Etienne Psaila, an accomplished author with over two decades of experience, has mastered the art of weaving words across various genres. His journey in the literary world has been marked by a diverse array of publications, demonstrating not only his versatility but also his deep understanding of different thematic landscapes. However, it's in the realm of automotive literature that Etienne truly combines his passions, seamlessly blending his enthusiasm for cars with his innate storytelling abilities.

Specializing in automotive and motorcycle books, Etienne brings to life the world of automobiles through his eloquent prose and an array of stunning, high-quality color photographs. His works are a tribute to the industry, capturing its evolution, technological advancements, and the sheer beauty of vehicles in a manner that is both informative and visually captivating.

A proud alumnus of the University of Malta, Etienne's academic background lays a solid foundation for his meticulous research and factual accuracy. His education has not only enriched his writing but has also fueled his career as a dedicated teacher. In the classroom, just as in his writing, Etienne strives to inspire, inform, and ignite a passion for learning.

As a teacher, Etienne harnesses his experience in writing to engage and educate, bringing the same level of dedication and excellence to his students as he does to his readers. His dual role as an educator and author makes him uniquely positioned to understand and convey complex concepts with clarity and ease, whether in the classroom or through the pages of his books.

Through his literary works, Etienne Psaila continues to leave an indelible mark on the world of automotive literature, captivating car enthusiasts and readers alike with his insightful perspectives and compelling narratives.
He can be contacted personally on etipsaila@gmail.com

Printed in the USA
CPSIA information can be obtained
at www.ICGtesting.com
LVHW091050021124
795328LV00003B/319